ISBN: 978-1-969991-99-8

www.DBEntertainmentllc.com

Contents

CHAPTER ONE

Episode 10: 3007

***

Audio Entry #01

"Today is June 30, 3007. The current temperature is 30 degrees Fahrenheit, and it is precipitating. My current location is at 44° 46' north, 73° 25' west, and I am here with my entire family. We have made a crash landing on the base of Mount Marcy due to severe damages to the right wing and rocket engines that came from a high-energy beam blast. This is what caused us to glide and dead-stick 262 miles away from The City of Neoterra as we tried to flee to safety. My family and I are trying to figure out how to get back to the International Space Station and escape Earth because it is no longer safe for us here. I have been removed from my position of power in The City of Neoterra because my reputation and name have been damaged due to my brother Mars. My own flesh, blood, and bones, Mars has deceived the people of The City of Neoterra and successfully turned them against me. I can't explain in full depth why he did this. All I know is Mars is filled with jealousy and envy. This is what caused the people of The City of Neoterra to lose their dependence and trust in me, and they are filled with anger. We do not want to be found by the citizens of The City of Neoterra because if my family and I are captured, we will be slaughtered. With the help of AERO, we can scramble signals to disguise our true location. We can use our signal to communicate with others, but it is not very strong, so we must use our signal wisely, so we do not get tracked down.

"Nonetheless, I can inform you that I am determined to find my way back to space. However, due to the condition of the spacecraft, I will have to try to find the materials I need to rebuild it from scratch. To reach outside space, we have no choice but to leave the spacecraft behind and get past Mount Marcy to reach a nearby city, where we can find the right equipment to rebuild the spacecraft. But if this reaches anyone. Please help. My youngest daughter, Sade, has become ill. I think everyone will be fine; however, I am worried about Sade. Due to the current cold condition we are living in right now, her body is unable to fight the cold. Her immune system is shutting down, and her sickness is getting worse. I must find the resources to create medication for her or at least rebuild her immune system. Plus, Sade has a mental condition, so she is unable to move on her own. Due to her inability to move on her own, it makes it very hard for us to move at the pace that we need to get where we need to go. We don't have the right equipment to assist her to be mobile.

This will make it difficult for her to survive this journey; however, we will continue to move through the snowstorm, and we will nurture her and take care of her by any means. At this point right now, I feel like I can't do enough to protect Sade and the entire family; however, I don't have time to live in regret. As a father, I want to bring my family to safety. I am their protector and provider and their keeper. If I don't do it, no one will."—Alchemist

Audio Entry #02

"What do you have when all you have is power? I came, I saw, and I conquered it all. However, in the end, I lost it all, and now I'm here, stranded and in fear for my life. Due to my circumstances, it's been days since I had anything to nourish my body. We were able to gather some food along the way; however, all the food we stored or picked up went first to my children (Sade, Hunter, and Carmen) and then to my wife, Scarlett, before I received anything to feed myself. I'm losing weight and energy; however, I know in the long run, it will be well worth it if we are able to reach our destination. It is still snowing heavily, and the temperature has dropped to 10 degrees Fahrenheit; hopefully, we can find shelter very soon."—

Alchemist

Later that day, Alchemist; his wife, Scarlett; his son, Hunter; and his older daughter, Carmen, carrying her younger sister, Sade, continued climbing through the mountain and the wilderness. As they passed through a small body of water rushing through a stream and a brook, Alchemist stopped to rehydrate his body. As Alchemist looked down to take a few sips from the running water, he noticed the physical changes that his body was going through. Alchemist looked down at his reflection, and he could see that his blue eyes were blood-shot red and his cheekbones were visible from dehydration and hunger. Also, he noticed that his peach fuzz had grown into a full-grown beard. Also, his once-dark blond comb-over fade had grown straight and long, as he was unable to find the time to groom himself. Once Alchemist gathered himself, the family approached a dark trail between groves of oak trees. Although it was still storming, the sun was out. But when they stepped onto the trail, the light became dim. It was very difficult to see where they were going, so Alchemist used the flash from his eye lenses as a source for light.

Midway along the dark trail, Alchemist and his family were stopped by a young white Lotus creature in their path. Everyone was stunned at first, but once they realized that the Lotus [1]didn't feel threatened, they (except Alchemist) slowly let their guard down. To the average eye, this Lotus appeared as innocent as a faceless child; however, it was a vicious creature. Slowly the Lotus approached Alchemist and his family. At that moment, Alchemist was in a defensive fighting position with his P9X rifle[2] in hand, prepared for the worst. Luckily for the family, a falling branch slammed to the ground and scared the Lotus; in it scattered into the forest.

---

[1] Lotus Creatures are a group of mutant/hybrid creatures that are fused with the DNA of an amphibian creature and human being.
[2] The P9X rifle is a futuristic military laser gun used by Alchemist and members of the United World Space Marshal.

***

It was a stormy night. Outside of Alchemist's palace, in the courtyard, a large group of people patiently waited for Apollo, the new leader of The City of Neoterra. Before he decided to speak to the people, he stood firm a few steps away from the edge of the balcony to soak in this moment of victory as the wind blew and hit his jet-black locs and caramel skin. Apollo steadily approached the end of the balcony that stood above the crowd of people. Once Apollo stepped foot on the balcony, he began his speech.

"Citizens! Gather around…how long before enough is enough? Is it a year, a decade, a century, a millennium? How long is enough? For generations, we have been divided and conquered; however, today is a new day. Although there is promise up ahead for us as citizens of this city, my question to you all is this: What needs to be done from this day forward? How can we make changes? How can we have our justice? How can we *earn* the respect that *we as people deserve?*

"As we stand here today, we stand here together free…free from all the oppression! Alchemist and his regime tried to trap us and kill us *one by one*; however, in the end, they failed, because we rose as one. One body. One mind. One soul. We fall as individuals; however, we rise as one. My brothers, my sisters: our day has come! We finally fought back, and now the power is in our hands! No longer will we walk blindly in a lie. No longer will we be misguided or deceived. No longer will we feel inferior. We break through those chains. Our minds are a lot stronger than they were yesterday. However, what do we need to do to move forward as a city?

"When you look back in history and see a nation that was able to conquer it all, although it was protected by a strong army, in the end, it was armed with knowledge. To move forward as a city, we need to empower and govern ourselves; we need to start building for our own, and we need to feed ourselves physically and mentally. For us to move forward and protect the next generation, we need to educate them. We must arm them with the knowledge of what we've been through and what we know so this will never happen to us again! No one will ever oppress us again! Prior to this day, active measures were taken against us. We were demoralized, and our minds were brainwashed to the point where we lost our perception. We lost our identity—we lost our purpose to defend our families, our city, and ourselves. However, today is a new day.

"Today is a new day because the man who has oppressed us is no longer in power. That man is Alchemist, and he is no longer welcome in this city. If he dares to set foot in this city, he will be captured, and he will be charged for the crimes he has committed. If he is captured, he will be charged with the crime of murder and all the evil schemes he used to leave us suppressed and trapped on this Earth. He will reap what he sowed. He can run in fear of facing the consequences of his actions, but he cannot escape his fate. He *should* be scared! And he *should* be running, because if he dares to set foot in this city, or if he is captured, he will fall as fast as he has risen. Today, enough is enough. We will come together as one! We will make the changes, have our justice, and *earn* the respect we all deserve. We will destroy everything Alchemist built. When we are finished, we will rebuild this city from the ashes *as one*! Let's go!"

***

At an altitude of approximately 220 miles, Dallas was at the International Space Station executive palace,[3] listening to his older brother Alchemist's plea for help.

"I can't believe Mars would do this. Why would he do this to him? Destroy everything that Alchemist built and then put Alchemist and his family's life in jeopardy for what. A coward—that's what he is. A coward! It goes to show you can't trust anyone, *even* the ones you call flesh and blood," said Dallas.

Dallas was pacing back and forth in his bedroom. In a circular motion, he was rubbing his right hand nervously, which was balled up in his left hand. Suddenly, he heard a knock at the front door. There were three calm knocks. Then the doorbell rang. A few seconds later, the knocks got louder and more violent. Before Dallas walked to the door, he looked at his surveillance monitors.

"Who is it?" asked Dallas.

"It's the UW space marshals! Open the door!" said Space Marshal #48201E-95.

---

[3] The International Space Station Executive Palace is the official residence and workplace of the president of the United World.

The space marshals had already flooded the Ayers property from the courtyard. But once Dallas opened the front door, there was a space marshal in every inch of the home, from the terrace, the foyer, the bedrooms, and the library to the lab room. The sergeant major of the army approached Dallas and issued him the warrant that allowed them to search and seize property from the home.

"What do you want? What are you guys doing?" asked Dallas.

"Under the Supreme Court of the United World, we have been granted permission to search and seize all property necessary for our investigation against Mr. Mars and Alchemist Ayers," said Space Marshal #48201E-95.

As the sergeant major was talking to Dallas, Dallas quickly turned his attention to the lab, where he watched other space marshals tearing apart everything in the room. Dallas watched in terror as they seized material from his chemistry lab, including everything from Dallas's beakers, test tubes, cylinders, Erlenmeyer flasks, DNA sequences, microscopes, and other machines to the biological materials used in his chemistry experiments, including bio elements, biomasses, tissues, cellular components, bodily fluids, and biological samples. They even seized Dallas's experimental animal farm as part of the criminal investigation. All Dallas's electronic storage devices were seized, including his computers and hard drives. They seized material from the greenhouse. The family's personal library of books was also part of the investigation. The library was a valuable collection that consisted of limited editions of scientific, academic, and mathematic books. It also consisted of Alchemist's personal scientific memoirs, journals, and special essays. The space marshal took all the books from Dallas's collections for the investigation. Even his pilot's memory chip was seized.

"Why are they taking all our stuff? None of this is Mars's," said Dallas.

"We need as much evidence as we can gather. It's standard procedure," said Space Marshal #48201E-95.

On the spur of the moment, Dallas's father, Zeno, walked into the room and embraced Dallas. "Hey, Dallas. How are you doing? Are you OK?" he asked.

"Yeah, I'm OK. Dad, what is going on? They are taking everything," said Dallas.

"I know, son. We have no choice but to let them do their job and take what they need to for the investigation…what happened to Mars? Where is he?" asked Zeno.

"I don't know. We got into a little scuffle, and then he broke away. That's when he told me everything I told you," said Dallas.

"We need to find him. Mars is in big trouble, and his life could be in danger. The court of justice is getting ready to place a warrant and send an army of space marshals to capture or kill Mars because of the chaos he has caused on Planet Earth," said Zeno.

"All right. Listen, Dad. Right now, he's the least of my concerns. Right now, I need to find a way to get to Earth and find Alchemist, because he's in real danger," said Dallas.

"Have you gotten in contact with Alchemist?" asked Zeno.

"Not quite. I receive an SOS that he recently sent out to space. I don't know where he is now, but I know he is with his family, and he's a few miles away from the city in the mountain area in search of tools to repair his damaged spacecraft," said Dallas.

"Is he OK?" asked Zeno.

"Yeah, he's OK," said Dallas.

"That's good to hear. I knew he would be OK, because Alchemist is a strong man. But you didn't hear from your brother?" asked Zeno.

"No! And he's not my brother or your son. That's your creation," said Dallas.

"You don't mean that," said Zeno.

"But I do. Listen: ever since Mars has been a part of this family, he has been nothing but trouble. I have no sympathy for Mars and his woes."

"I understand why you feel the way you do; however, we still have to protect him and be at his side, no matter what he did. He's still a part of our family," said Zeno.

"I don't understand why you are so worried about Mars instead of Alchemist. He's your real son, and he's in real danger right now. But the only person who is on your mind right now is Mars. I don't understand. Besides, if not for Mars, Alchemist wouldn't be in the predicament he's in right now," said Dallas.

"I know; I know. But listen to me, Dallas; if you see or hear from Mars, let me know. This is very important," said Zeno.

"Whatever…anyway, when will I be able to use the lab again?

I need the lab to help find Alchemist."

"Due to the investigation, you will lose access to the lab. I don't know how long you will lose access, but if you really need one, Alchemist has a lab in an undisclosed location. If you want to use the lab, I can put the directions into your navigation system, and it will take you where you need to be," said Zeno.

"That's perfect; just put in the directions into Pilot's system, and I'll be on my way."

"All right. Pilot, come here. Here are the coordinates. As soon as you get to the lab, call me. And Dallas, be safe…"

"All right, Pops. Thanks," said Dallas.

With the help of Pilot, Dallas drove across the International Space Station in search of Alchemist's secret laboratory to help save his brother, Alchemist, and his family.

***

Sitting in a temple in the city of Utopia, Beta of the universe, Queen Nebula was praying to the Alpha god. The soon-to-be Omega of Planet Nibiru, Iris, entered the same temple to discuss the security of Planet Nibiru. Iris didn't hesitate or beat around the bush; he was straightforward in what he wanted to talk to her about.

"Where are the rest of the humans?"

"I don't know."

"Don't play naïve; I already know that Zeno sent an army of men to try to take over Nibiru. He's found our weakness and started a plague against us, and now we are vulnerable. I know how an army leader thinks. As commander in chief of the Titans[4] and Amazons,[5] if I were leading an army of my soldiers against an enemy, and I knew that their army and their people were in a state of vulnerability, the first move I would make is to attack and start a war against us. We need to prepare for war. A holy war is at hand, and we need the gods' best soldiers to fight and defend this land. However, we need to know where they are and where they are going to be. So, I am going

---

[4] A Titan is the gender for a male Nibirian.

[5] An Amazon is the gender for a female Nibirian.

to ask once again—where are the humans? Where are the space marshals?" asked Iris.

"Honestly, I do not know where the humans are or where they are going to be. Besides, I don't understand why you see them as a threat because they are nothing but innocent creatures," said Queen Nebula.

"Are you being sympathetic to the humans? Have no sympathy for the devil or his serpents! Your heart has blinded your mind to the point where you have confused humans with saints. Humans are not innocent creatures—they are savages, built for war and destruction, and soon we will be the victims of their wickedness if we let them do whatever they want. They don't deserve to be here on Planet Nibiru," said Iris.

"But the humans are the goddesses' creatures, and the goddesses have a purpose for all of their creation. We cannot destroy out of fear what the goddesses have created. There is nothing to fear because the goddesses would never allow anything that isn't meant to be in the city of Utopia," said Queen Nebula.

"You know, I always question whether or not you really have the soul of Nebula, because I don't think you're fit to be a Beta god. You welcomed these demons and led them to our land. And now you defend them? Are you serious? They are demons! A bunch of parasites—worthless creatures made from nothing but dust!

"Explain to me, Iris, why do you feel so deep in your heart that the humans do not deserve to be here on Planet Nibiru."

"They're built for war and destruction. Humans can't live without living in dysfunction and can't live in eternal peace like Planet Nibiri. You would hope once they arrived on this land, they would leave their primitive desires behind, but this is not the case. The *Homo sapiens* are beautiful. It is their soul that makes them conscious and gives them morality. But without the soul, they are nothing. It is their flesh that controls all their desires. It is pathetic how fragile the gods had made humans. They are weakened by their emotions with so much lust for power and control that they are willing to risk it all and go against the gods for what they believe is their divine right. And it's understandable; here in Planet Nibiru, it's a blessing to be able to live in this planet because everyone is free from suffering and is immortal. However, on Planet Earth they are dealing with a lot of suffering.

That's why they are willing to travel light-years away to come here and take away from us what the gods have blessed us with because humans have been cursed and they are subject to death. The gods have punished them for their wrongdoing, but now they are trying to take matters into their own hands. They are trying to escape all their wrongdoings on their planet and strive for things they know they don't deserve.

"Do you think they pose a threat?" asked Queen Nebula.

"Absolutely not. They are humans, and we are the chosen ones! However, we must not take the humans lightly. That's why we have the best Amazon and Titan warriors behind us to fight this war."

"If we are the chosen ones, then why are you so worried about them?" asked Queen Nebula.

"For the past millennium, humans have demolished and drained the source of life from other planets in search of new hosts to keep them alive. And now they are trying to find refuge in the city of Utopia. We must stop them and take back our planet before it's too late for us as chosen ones," said Iris.

"Iris, you are overthinking it. There is nothing to worry about. They are humans. If there was something to worry about, I would have informed you, but there isn't. You need to calm down," said Queen Nebula.

"That's it. Enough is enough. From this day, as the Omega of Nibiru, I declare that you will be marked as a traitor to Planet Nibiru and its people. You have turned your back on the gods and can't ever be forgiven. You are no longer welcome in the city of Utopia and Planet Nibiru. You sold us to demons for your passion for the human, Alchemist. I'm the Omega of Planet Nibiru, and we share the throne. The gods chose me for my strong genes and for being the most powerful and smartest warrior in the galaxy; however, it seems that you traded my love for you for a human! I love you with all my heart; however, I believe the reason you abandoned me, and our people is because we have been unable to conceive a child. My love for you can never be matched, but because we were never able to conceive a child, you betrayed me and the people. There is no reason for me to keep you here. Leave—and never come back."

"That's fine, because Alchemist was able to achieve what you were never able to achieve: a child. I was able to build a relationship with Alchemist. We are bonded by our love, and the child will soon be welcomed to the universe. You may be the most powerful and smartest warrior in the galaxy; however, Alchemist is really the one who was chosen to be second to the throne. Soon we will breed this child, who will share the spirit and the breath of the gods. And once it appears on the surface of the universe, from the ashes of the city of Utopia, he will have the power to create a new heaven and new planet, which will be devoid of all sins and suffering. Good-bye, Iris. I'm going searching for Alchemist, and when I return, you will be long gone. Your judgment day will come very soon, and you will be punished for all your evil and wicked ways. And that human everyone so fears will be wearing your crown and sitting on the throne," said Queen Nebula.

"Whatever. Leave, and never come back," said Iris.

Queen Nebula got up from her seat, walked out of the temple, and left Iris in the temple alone as she began to travel away from Planet Nibiru.

***

At the army base, a group of space marshals were traveling back to the International Space Station from their last tour as part of President Zeno's mission to send over a hundred thousand space marshals back home. After serving his time as space marshal for the UWSM[6] (the United World Space Marshals), war veteran Space Marshal #2339IE-5 Calvin Davis finally returned home.

Calvin was relieved that he was finally home; however, his circumstances back home forced him to face new challenges, which most veterans must face when returning home. While traveling through the intergalactic space, Calvin Davis enjoyed himself, shuffling through the television channels. He caught the tail of an NSFL (National Space Football League) playoff game: Station 66 Steelers versus Station 51 Space Raiders. They were broadcasting live from the SILO (Stadium International Lunar Olympics) Stadium in the space station, and the score was Steelers 10 and Raiders 48, with two minutes left in the third quarter. Although the Steelers had no chance of a comeback, Calvin was willing to wait it out because the

---

[6] UWSM stands for the United World Space Marshal. The United World Space Marshal is the space army for the United World/International Space Station.

Steelers were his favorite team. Calvin was hungry, so during the commercials he grabbed something from his Space Snacks® vending machine. In a matter of seconds, Calvin was able to produce a full-course meal with only three small capsules placed in the microwave. He made clam chowder, Caesar salad, and a sextuple-hex tuple bacon cheeseburger, and for dessert, Space Snacks® Space Cookies®. Once he had all the essentials he needed, Calvin returned to his station, where he continued to watch the game. But during the fourth quarter, the score increased to 10–55, and he lost interest and again started shuffling through the television channels.

(TV static sound effect as he changes the channel.)

"I am E-ton 9548. You are now watching NSBA (National Space Basketball League) on Station 46."

(TV static sound effect as he changes the channel.)

"The winner of Miss Universe is Miss Europa. Elizabeth is the first extraterrestrial to win Miss Universe."

(TV static sound effect as he changes the channel.)

It took a while before Calvin came across a documentary that grabbed his interest. The documentary was about the trials and tribulations of the rise and fall of Zeno Ayers. Another space marshal named Shane walked near Calvin's station area and caught a glimpse of what he was watching. "What's up, man? What are you watching?"

"It's a documentary about the creator of the International Space Station".

Zeno Ayers was the president of United World and a business mogul. The Ayers family originated in Weston, Massachusetts. Zeno came from a wealthy and elite family that included scientists, engineers, bankers, and investors. According to *Forbes*, they were estimated to be worth over some quadrillion dollars. The Ayers family established their wealth in Wall Street investments and were part of creating some revolutionary-technology project; however, one of the family's biggest investments was war bonds. The last time the United States issued war bonds was during World War II, but during the time of World War III, places like the late great America were in extreme debt and needed funds to defend themselves in war. The Ayers family already possessed a significant fortune before the start of World War III (2020–27), and the family had gained preeminence in the bullion trade by this time. The family made money from war bonds with multiple nations during World War III. They single-handedly financed World War III for both America and their continental allies, producing over $500 billion in revenue. In the process, the family developed a network of agents, shippers, and couriers to transport gold across war-torn Europe."

"Wow, that's crazy. Who would have thought they could create a nation in space from war? That's amazing. That's why I always question war. Why do we go to war when we have all this pussy back home (looking at a photo of a nude extraterrestrial female)? It's because we're fighting a rich man's war. It's always been about the money," said Shane.

Calvin cut off Shane from his short rambling rant and focused again on the documentary. "Yo, Shane, can you shut the fuck up? I'm trying to watch this shit."

"My bad. I'll leave you alone, douchebag."

The family network was also to provide the Ayerses with political and financial information ahead of their peers, giving them an advantage in the markets. As the Ayers family continued to establish themselves within the market, they were raising their children to take over the family's empire, preparing to hand over the wealth to the younger generation. They successfully kept control of their banks in family hands, which allowed them to maintain full secrecy about the size of their fortune. The Ayerses successfully kept the fortune in the family with carefully arranged marriages, often between first or second cousins (like royal intermarriage).

"Out of all the children in the family, Zeno showed the most potential. At the age of thirteen, the prodigy Zeno began to build the technology that would allow humans to live in space. By the time he was twenty-three years old and had graduated from Massachusetts Institute of Technology in Cambridge, he'd completed a small prototype of the interstellar ark. He presented the idea to his father, mother, uncles, and aunts, and they invested three-point-five trillion dollars to build a space ark. With the help of NASA, Zeno was able to create the first interstellar ark," said a commentator.

The construction took place in space near the International Space Station, and it took ten years (2040–50). By the time humans finally lived in space, the Ayers family had already established a universal space bank, where they had multiple assets in stocks, bonds, and debts around the universe. This was the biggest revolution in technology since the digital revolution.

"The Ayers family moved over seven hundred trillion dollars of their own money from all over the world into space and started their own businesses in space, such as oil products, agricultural products, steel production, food-processing industries, water, metals, minerals mining, banking, energy, and farming, as well as being producers of essentials for the atmosphere for the International Space Station such as nitrogen, oxygen, water vapor, argon, carbon dioxide, and much more," said another commentator.

However, controversy surrounded the Ayers family when the friction between Pluto and Zeno was publicized. "It dated back to when human civilization was still on Planet Earth, and Pluto was creating the space ark. Pluto Varyn graduated from MIT in engineering the same year as Zeno. The two families have been in competition since then. Zeno stole the blueprint for the International Space Station (the space ark), and now Pluto's lifelong mission was to take over Zeno's empire and cement his family name as the elite family. Pluto held a grudge against Zeno because he stole blueprints to the space ark and took all the credit for it. It is alleged that Zeno was recognized as the sole inventor, although it was the group of scientists and engineers that he assembled who help him create the first interstellar ark," said a commentator.

"Pluto used to be part of that elite group of scientists and engineers, but he didn't get any credit or patent because Zeno had the resources to fund the project and gather his own staff at a good price. Pluto was among those many in his group. Zeno was able to establish and launch the first-ever colony of people to live in space. The International Space Station was the only interstellar ark before Pluto was able to build his own," said the second commentator.

"Once Pluto was able to establish his own ark and build his own society and army, he took the opportunity to wage war against his predecessor. He was able to conduct the research that helped him develop the AI for the force of humanoid robots that were used to attack and start one of many space wars against the International Space Station," said the third commentator.

***

Later that day, Zeno traveled from the lab to the International Court of Justice based in the Peace Palace for an emergency special session. Within twenty-four hours, Zeno, who was the president of the general assembly, and the rest of the general committee (the main organs of the United World: the vice president of the United World and the six main committees) were in attendance. General-assembly observers who were granted observer status due to their contributions or because they were suppliers of a good or a service or owned war bonds for the United World and the International Space Station, were in attendances as well. The general assembly observers consisted of the wealthiest and most powerful men and women in the International Space Station. Whether it was the army commander, space merchants, scientists, armament manufacturers, asteroid miners, or water cartels, they were in that room waiting to figure out the mission and agenda of this general debate. The meeting was called at the request of the Security Council to settle multiple legal disputes submitted by states and to find resolutions and decisions for considerations that had recently emerged. And as the president of the general assembly, Zeno was going to open up and preside over the session for the members of the United World.

"The informal plenary meeting of the general assembly is now called to order. I now give the floor to the secretary-general, His Excellency James Aesop," said Zeno.

"Fellow delegates, ladies and gentlemen, *mesdames et messieurs, damas y caballeros, nǚshìmen xiānshēngmen, dam i gospoda*: Welcome to the Peace Palace. Thousands of years ago, the entire world was in total chaos. World War II had ended, Hitler and his Nazi regime were conquered, Japan was defeated, and the Soviet Union was disabled. But with so much bloodshed and hostility, there was a good potential that World War Three would happen. However, once the United Nation emerged as an international system, people all over the world with different beliefs were able to settle their differences and come together. The United Nation eliminated the division that was building and brought democracy to the entire world. We were originated as a system that would bring order and enforce the dignity and equal worth of all people. The United Nation enforced international law, security, economic development, social progress, and human rights

for everyone in and around the world.

"The United Nation originally started off with fifty-eight

members out of the hundred and ninety-six countries and two

hundred and forty-eight nationalities scattered across the world. Once

the United Nation became an official system for cooperation among

allies all around the world, we were able to bring peace and liberty,

improve living conditions, and bring diplomacy and international

order to the world. Without the principles and the influence that the

United Nation established, countries and people all over the universe

wouldn't be able to practice democracy. The United Nation and the

system that drives it are essential tools to bringing order to the

universe; however, this took centuries of work. Today, as we are

gathered here at the International Space Station, we stand here as one

nation. Today is 3007, and if you look around, you can see how we

were able to take their vision to the next level. We overcame the

difficulty, and we embraced the challenge, and in our quest for

knowledge and progress, we were determined to bring the world to

outer space. And this all started from a vision that started in 1969

when the late great Americans became the first on the moon. "With the help of Zeno and his family, we were able to rebuild the International Space Station and move to space. Our main goal was to unify the entire universe and form this super nation, and we succeeded. After we rebuilt the International Space Station into an interstellar ark; we brought Elites from all over the world to form the United World. Slowly and surely, we were able to move forward and implement a system of universal rules and norms. We were able to get past the desire for power, dominion, and supremacy and consider a future where there is peace and harmony while coming together. We were able to create a universal currency and build the strongest army and government body ever known in history. However, although we were able to establish this institution that would work for the good of humankind, for centuries we have continued to fail to practice the principles of the United Nation. Because of our wrongdoing, our foundation is breaking apart. I don't know how much longer we can live in the International Space Station, because the infrastructure of the station is falling apart. It's a shame that despite all our technological advances, we are still dealing with environmental issues such as waste disposal or pollution in ecosystems. This is what is causing climate changes and natural resource depletion. The chemical contamination is breaking down our air, our crops, our soil, and our water. And what makes matters worse is that our population has surpassed the capacity of the space station. We have overpopulation, scarcity in water, and space pollution".

"This is what forced our people to make Planet X their new host planet which result in them getting captured and enslaved by the people of Planet Nibiru. We must find a way to bring them back home whether it's through negotiation or war. The same goes for the Republic of Pluto; there have been individuals at the Republic of Pluto who have been caught using inferior technology like drones and androids built with super artificial intelligence to start biological warfare against innocent people. We must counter these barbaric attacks one way or another. Moving forward as a nation, we must stay true to the model that has brought us to where we are today despite the status of the universe. Instead of embracing our deepest fears, we must believe that things will get better for humanity. The United World was founded by a group of men and women who had a vision of a peaceful universe where all nations were able to uphold basic laws and needs for humanity. As a nation, we must take the responsibility to maintain international order while practicing peace rather than going to war by using the principles from the space law and international declarations. That's what we are here to do today".

"As we construct our sustainable-development goals for the next fifty years, we will show the strength of the United World by refining the universal system that has been working for centuries. The only way we can resolve these issues is if we let go of our old differences and find a common ground where we can bring democracy, peace, and international order to all people. As secretary-general of the United World, I'm aware of the dangers ahead; however, we must remain content. We cannot return to the old ways of dealing with our issues with one another. We must remain content. We can't fight fire with fire. We must fight fire with water and use effective methods to achieve what needs to be accomplished. That's the only way we can move forward as a whole. We have created standards that are unrealistic; however, we created a model we should live by. 'Cause if we meet fear with fear, or meet hate with hate, we continue a cycle of self-destruction that can last forever. History shows that this is not the way to go; all the fallen empires—ancient

Rome or the late great United States of America—tried to use divide and conquer but met their fate. They all failed".

"John F. Kennedy said, 'Mankind must put an end to war before war puts an end to mankind.' If we want to move forward, we must be diplomatic as a tool to work together more effectively. If we cannot come together as one and strive for freedom and peace for all, we will suffer the consequences of those who tried to meet fire with fire. For the sake of the universe, we can't go backward. We must come together and find interests and principles that are universal, which are the cornerstones of the United World. I believe that if we all stay together and choose collaboration over conflict, we can bridge all our differences in this interchangeable universe. As I stand in front of you today, I ask you all to let us carry out what helped build this institute. If we can move forward and bring faith and hope into the future, the future will be brighter, not just for us as a nation, but also for everyone in the universe. In such a dark universe as the one we all live in, we were able to find the light through it all. Progress is real: we were able to come together fight for humanity, society, and create universal order. We fought against corruption, hunger, and diseases that we never thought we'd be able to cure and save billions of lives.

However, our mission as a nation is far from complete. Today we find ourselves being pulled back into a darker and more disordered universe than ever before. We are in a state of emergency and this must be addressed if we want to move forward as a nation. Thank you very much."

***

After Mars returned to space from Earth, his first destination was Pluto's space station. Standing in front of Pluto's family palace was Mars. He buzzed to enter. Once he confirmed who he was, he was granted entrance. Adakis greeted him at the front door and welcomed him. Mars entered the greeting room and was introduced to the Pluto family. There was the prince, Adakis Varyn. There was the first lady, Alexandria Varyn. There was Angel, who was a close friend of the family. And there was Pluto Varyn, who was leader of the Pluto nation.

"How are you doing? Welcome. Do you need anything to drink?" asked Adakis.

"No, I'm fine. Thank you."

"All right, then. Wait right here; my father should be coming down from his room in a few minutes."

Pluto entered the room. And as he walked down the steps, he quickly resumed the last conversation he'd had with Mars. "Did you complete the mission that was assigned to you?"

"Yes."

"Good. Good. Good—very good. Once your father realizes what has happened, he will feel the vengeance of his lost and forgotten son. What type of father would create and give birth to a son and then send him away to fend for himself? You were your father's second son. But for some reason, he didn't see it that way. See, at his birth, I chose my son Adakis to revive the Pluto dynasty to its glory years. My son witnessed and learned from my mistakes in the past. This will help to prepare him to lead this nation after I die or abdicate. Adakis is the prince of the dynasty, and I am proud to say that he will be greater than I have ever been. Your father feared that his children would become greater than he was. You should be the prince of your father's empire. But that's why I'm here. I will give you the opportunity to get the power you deserve, and the next time your father stands in the presence of his son, you will be recognized as a ruler in control of the entire universe, and you will be glorified as such. And when he kneels before his son, he will give an account for all ill will he displayed toward you. Anyway, this is not the time or the place to have this discussion. Come with me, my son, so we can discuss our plans."

"Yes, sir," said Mars.

***

Audio Entry #03

"Hello, this is Alchemist, and this is my audio entry number three. My family and I are continuing our long journey through Mount Marcy. The entire surface of our path is still covered in snow, but we were able to push our way to create a path to our destination. We have passed a couple of abandoned bridges, brooks, creeks, and damaged, deserted, unpaved roads before reaching another trail. The trail was very steep and difficult to maneuver; however, we were able to make our way through. We now just passed a new trail, which led us behind a ranch that was standing on sixty-six acres of abandoned agricultural land. This land seemed mythical because while everything outside of it was covered in snow, the land was filled with natural resources. This is perfect because this land has all the essentials that I need to restock our supplies for the journey and help Sade. The land has many gardens, a variety of domestic animals, and other crops that we can use, so this is great."

At this point, all the Ayerses' devices have signals and are back on the grid. This allowed Dallas to connect directly to Alchemist.

"Dallas to Alchemist, can you read me? Dallas to Alchemist, can you read me?"

"Dallas? Is that you?'

"Yes. It's me. Are you all right?

"Thank god. Yes, I am all right."

"Where are you?"

"I just reached this large farmland a couple of miles from where my spacecraft had crash landed. Did you receive my SOS?

"Yes, I did."

"OK, then you know what's going on. I am currently standing next to what seems to be the main ranch. I don't know if someone lives here or owns the land. I don't want to intrude without being welcome. What I want you to do is to look up some information on my current location to see if anyone lives here. In the meantime, I will send you my daughter Sade's vital signs, blood, and urine samples to analyze and figure out her macromolecule levels and see if she's dealing with anything else. She's sick. I need to know what's going on with her body. There are a lot of resources here that I can use to my advantage if we need to come up with a formula for medication," said Alchemist.

"Confirm. Send me her vitals, blood, and urine, and I'll send you information on your current location," said Dallas.

"All right."

To send Sade's vitals and fluids, Alchemist used the digital component of his garment, which was equipped with a heart-rate monitor and a quantum teleportation device. A few minutes later after sending the specimen, Dallas returned with the result.

"Dallas, what do you have for me?"

"I have some information on your current location. The land you are standing on right now is owned by a man who goes by the name Abe Smith. From the data that I was able to gather, it is said that he is a 1,113-year-old blue-collar man who has been farming just as long as he's owned the land, which is a thousand years. It is said that he became immortal after finding the elixir for eternal life. He went in exile after he and his family were threatened, as many people wanted to get their hands on what was called 'the gate to heaven and immortality.' After 2050, Abe Smith disappeared from the face of the earth following a period when he and his family were constantly harassed. There is no indication that he is deceased, so I assume that he still alive. It is said that he may have returned to his farmland after giving up the fountain of youth. I suggest that you see if he's there, and if he is there, find out more about him.

"As far as the status of Sade—Sade's macromolecule levels are low in all three categories: carbohydrates, fats, and proteins. Sade is deficient. She needs to synthesize or ingest those ingredients to rebuild her immune system. Also..."

Alchemist lost the signal.

In the meantime, Sade's immune system was failing, and she was having complications. They had to wait a couple of hours before Alchemist was able to get the full results of the test. To boost Sade's immune system, Alchemist needed basic IV solutions to refuel her with fluids and electrolytes. Alchemist began walking the acres of the land, and as he and his family continued walking, they found a field of fruits with an elderly man sitting in a rocking chair nearby. As Alchemist approached the elderly man, he began to smile and wave with happiness as they got closer.

"Greetings, my friends. My name is Abe Smith; how can I help you?"

"Hello. My name is Alchemist, and this is my family. The one on your left is my son, Hunter. The girls on the right are my two daughters—Carmen, the older one, and Sade, the younger. And finally, this is my beautiful wife, Scarlett."

"Pleased to meet you all. So how can I help? It looks like you guys have been on your feet for days," asked Abe.

"Yes. We need some help and time to get our feet back up under ourselves," said Alchemist.

"And I am here to help you guys. Here, come with me," said Abe.

As they started walking back to the main ranch, Sade began coughing violently. Abe was concerned, so he asked if she was OK.

"No; she's actually dealing with a sickness and needs to be treated with some essentials that can be found in your garden."

"All right. I will walk you guys back to my ranch, and then Alchemist and I will head back to my garden and grab something that will fill your bellies and help Sade with her sickness. Sound good?" asked Abe.

"Sounds good."

"All right then. Let's go," Abe said.

When they reached the ranch, Abe gave each member of the family a room to rest in to restore their energy. Afterward, Alchemist and Abe went back outside in search of food for the entire family. As Alchemist and Abe walked toward the garden, as an icebreaker, Alchemist asked Abe if he had children of his own.

"Yes, I do. I have two sons, and I have two daughters; however, I haven't seen them for nine hundred and fifty-seven years," said Abe.

"Nine hundred and fifty-seven years?"

"Yes, nine hundred and fifty-seven years. See, what you may not know is that I am one thousand one hundred and thirteen years old and have been farming on this land for one thousand years. In my last stage—or what I thought was my last stage of my life—I started to feel like I was dying. As I was getting older and coming close to my death, my fear of death gave me the desire to find the fountain of youth. I searched and did countless experiments for years until I finally found the solution for the fountain of youth. With the fountain of youth, I was given the ability to rejuvenate and revive myself and remain young and cheat death. Now I am here today telling you my story," said Abe.

"Wow. But why nine hundred and fifty-seven years? What happened?"

"Well, it was before the 2050 apocalypse when I found the solution for the fountain of youth. When I found the solution, I drank it and gave it to my children and wife, and we kept this secret about the fountain for two hundred years. Then we came out and told our story. At first, we were praised and treated like royalty for the discovery; however, after 2050, when I broke my promise that I would not mass-produce the fountain of youth, my sanity was threatened. It was the beginning of the apocalypse in 2050 when I had no choice but to send my two sons, my daughter, and my wife to space to get away from the chaos that was happening on Earth. I sent my family to space and told them to never come back for our sanity."

"Your sanity? Why would you be in fear of losing your sanity when you are immortal?" asked Alchemist.

"I tell people, 'It may seem like a blessing that I am able to live for so long; however, it is not as pleasing as it seems.' I tell people that life is precious; however, it is not worth living forever. I tell people that even though you're immune to death, you are not immune to the feelings that everyone must face every day. I tell them that immortality brought sorrow to my heart. There will never be peace on Earth. I have seen constant battles and war; I have seen massive starvation and other senseless acts. That is the nature of life. I regret finding the fountain of youth because when you live for so long, you will see things that you wish never to see. I was murdered multiple times. I saw my family tortured and then come back to life. That is why my soul will be in constant battle. My soul is trapped because it is constantly trying to escape the flesh. I feel like I am being punished by God because I tried to cheat death. I wanted to be free from fear; however, it resulted in the entrapment of my soul," said Abe.

"Do you still have the fountain of youth?" asked Alchemist.

"I know where it is; however, I will never give that information to you or anyone because I know it's dangerous. Even though I am immune from dying, I still see the imperfection of life. Although I can't die, I have witnessed death and generations of greed and corruption. I tried to keep this away from anyone other than myself to protect humanity, to protect them from what I am forced to see every day. However, other people wanted eternal life as well. I told them to reach for eternal life but not on Earth. However, my sanity was threatened. My mind was scarred so much that I gave up and gave them the fountain of youth," said Abe.

"Who did you give it to?" said Alchemist.

"I don't know who they were; maybe some beings of higher intelligence. However, whoever they were, they were very wealthy and very powerful, and if they used the elixir for eternal life, they are probably now in control of something big and corrupt. No one can stop them because they know the outcome of everything, since nothing changes, everything stays the same, and history always repeats itself..." said Abe.

Alchemist and Abe reached the garden. Abe instructed Alchemist to grab a bucket and fill it up with short stalks of raw sugar cane. They found a salt solution inside a nearby lake after using a simple distillation. Alchemist gathered animal protein to help refuel his family and himself. There was a variety of domesticated animals: cattle, chickens, horses, pigs, goats, and sheep. From a single goat, Alchemist used the skin for clothing and the flesh for meat for his children and wife. Alchemist was able to produce cheese and yogurt from scratch from the goat's milk. From the cattle, they used the liver and beef as a source of energy as well. From the spice garden, Alchemist filled baskets with pungent spices—curry, ginger, cilantro, rosemary, thyme, cloves, nutmeg, and other spices—to help give their food some flavor.

Once Alchemist and Abe gathered everything they needed, they returned to the ranch and gave Sade the basic IV solutions inside a plastic bottle that was filled with a sterile water compound made from salt, sugar, and additional minerals. After everyone was fed and once Sade was in stable condition, Scarlett and Carmen bathed her in berry juices and herbs.

"Where's the towel?" asked Carmen.

"It's in the other room," said Scarlett.

"Let me go get it."

As Carmen walked into the other room, Scarlett slowly dipped Sade's head under water. As it became difficult for her to breathe, Sade began to squirm for air. Scarlett held her under water until Scarlett heard Carmen's footsteps. Once Scarlett pulled Sade from out under the water, Sade began coughing dramatically. Scarlett began patting her on the back to help her breathe again.

"Oh my God. What's wrong? Is she OK?" asked Carmen.

"Yeah, she is OK. She just got a little water stuck in her nose."

Carmen paid that moment no mind. After they finished bathing her, Scarlett tied her hair into a ponytail and then put back Sade's favorite space suit dress before she fell asleep. She then laid her down in her bed.

***

Audio Entry #04

"Before me and my family were getting ready to leave the farmland, I asked Abe if he had the right equipment to repair the Space Arc. He did not have the right tools at hand; however, he knew where to find them. Abe gave me the directions, and we are now prepared to move to the next destination. However, as we are getting ready to set up the horse and carriage, Sade's body began to reject the food and medication. Her immune system was shutting down. At first, I noticed Sade mimicked the signs of a common cold: mild coughing and a runny nose. However, now Sade's condition has gotten worse. I noticed definite symptoms from Sade. Her fingers and toes trembled, and her mental state has changed. Her skin was blue and gray, and her veins were showing. I checked her body's temperature, and it had dropped to 35.6 degrees Fahrenheit. And she was now developing hypothermia. I mistook her symptoms." Suddenly Dallas reconnected with Alchemist.

"Alchemist? Are you there?"

"Bad news. According to the samples you sent me, there is a bio accumulative toxicant substance in Sade's system. The tests revealed that she had a blood-mercury level of 192 micrograms per liter, double the toxic threshold in the human body. This substance can be very deadly if it is not treated. Is she all right?"

"No, she's showing the symptoms right now. Were you able to come up with any treatment while we were disconnected? I need something right now to help her before it's too late."

"Yes, don't worry; I am going to transport the compound to help her right now. During the time you were offline, I was able to produce a compound. I will teleport a duplicate of the compound just in case the first one I send is destroyed in the process," said Dallas.

"Hurry up."

While Dallas transported the compound to Alchemist, Scarlett held Sade. When Alchemist was getting ready to inject the compound into Sade, Scarlett wouldn't let her go.

"What are you doing? Let her go. I won't let anything happen to her; I promise."

Scarlett handed Sade over to Alchemist, and he injected the compound into Sade's body. A few minutes later, her condition improved. Her macromolecule levels were restored, her body temperature rose to normal, and her breathing and her heart rate were both normal. Once Alchemist realized that she was going to be OK, he wrapped his arms to keep her warm as he rocked her back and forth as he shed tears of sorrow. Once Sade was in stable condition and knew it was safe to travel, Alchemist and the family saddled the horse; the rain ceased, and the sun rose. And under the sunrise, Sade smiled once again. At this moment, Alchemist finally was able to find joy in his wounded child, Sade.

***

Citizens of the United World constantly praised their soldiers and honored them for their service; however, they never saw the things that they faced every day. Most people envisioned the soldiers being greeted with love and affection by close ones. However, this wasn't the case for some space marshals including Calvin's family (his wife and child). Calvin Davis and the space marshals finally reached the International Space Station. Some of the space marshals were greeted by family members, but like the rest of the space marshals; Calvin Davis had no one there to greet him. It had been very hard to keep in contact with his daughter and wife, who were homeless and constantly moving from shelter to shelter.

While Calvin was fighting in the space war, he was light-years away from his family, and there was no way he could stay connected with his family—no satellite or hologram—so they were not informed that he had returned. However, Calvin took a bus to the Nation's capital in Station 13 because he was determined to find his wife and daughter. Calvin chose Station 13[7] because the likelihood of finding them there was higher. This was where they lived before they

---

[7] Station 13 is a city in the International Space Station.

went homeless and the last place where he'd seen his family. In the meantime, Calvin tried to reestablish himself in the city; there had been a lot of changes since the last time he was in the International Space Station. He roamed the city and saw the changes.

The last time Calvin was living in the International Space Station, humans were the only ones who populated the interstellar ship. However, Calvin noticed things had changed. Synthetic humans, robots, and extraterrestrials were living and socializing among one another. As Calvin looked at the horizon, he also witnessed all the new buildings, replicas of the universe put into the space ark. Calvin was amazed by all the wonders of the universe and other planets in the galaxy at the same time.

In the meantime, he was able to pass most of the wonders of the International Space Station. The best creators in the universe were involved with constructing the International Space Station. With the collaboration of architects, scientists, and engineers, the International Space Station was structured in a Bernal sphere design.

And in a few miles' radius, a person could travel from Central Park to Hong Kong, fly through the Golden Gate Bridge or Brooklyn Bridge, see landmarks such as the Christ the Redeemer and the Eiffel Tower, or sit in the elements of Antarctic, Hawaii, or Southern California and enjoy many other wonders of Planet Earth in a matter of minutes. These top architects, scientists, and engineers were able to create their own ecosystem. This allowed them to produce and create an interior of the International Space Station consisting of ten different synthetic natural habitats—mountains, coniferous, deciduous, or tropical rain forests, along with wetlands, deserts, Iceland, Greenland, agriculture surfaces, and urban areas. From where he stood, he had a view of the synthetic forests with genetically engineered trees, hills, mountains, rivers, and oceans. With the weather-modification technology, Calvin saw the cloud banks and the Aurora Borealis blending in the beautiful blue sky and felt the air and gravity the same as on Earth. And through the window, Calvin had a clear view of space, with the sun, Earth, and other planets in the background. Calvin was truly amazed, but once he caught himself getting distracted, he returned to his focus: finding his daughter and wife.

***

"Live from Cyber News headquarters, today is 1070-09-20 ∇ 21:00 lunar standard time. Hello. My name is Michelle Waters, and you are watching Cyber News.

"Top of the news today: At ∇ 21:35, we will discuss whether or not having a personal robot is safe or a weapon for mass destruction. With the murder rate between humans and robots rising due to system malfunctions, manufacturers and officials are coming together to figure out whether the artificial intelligence built into these machines is suitable for household usage.

"But in today's news, aircraft crashes are at an all-time high. With close to a hundred thousand people dying in the past year in crashes, International Space Station officials are planning to create new precautions to prevent this issue of aircraft crashes. International Space Station police data reveals that the number-one reason aircraft accidents have risen is the increase of drinking and flying. The International Space Station police data also reveals that another reason aircraft fatality have increased is because people simply can't handle flying aircrafts. The International Space Station police are planning to lower the speed and the altitude at which people can fly their aircrafts to lower the chance of fatality.

"In other news, the trillion-dollar company Parashoes Manufacturing has recalled over five hundred thousand of its Rocket Boots today due to multiple reports of the flying boots exploding, which has caused thousands of people to lose limbs. The president of Parashoes, Inc., made a statement today: 'As one of the most recognized businesses in the universe, we pride ourselves as a corporation in creating the best and safest products in the transportation market. However, as president of Parashoes, Inc., I will assure you that our staff is going to put full effort into restoring the lives that were affected by this huge setback for our company. It will take a while to fix the damage that was done; however, our entire staff is willing to face the challenge ahead.'

"This setback will cost them billions of dollars in damages; however, the company hopes that they will be able to fix this issue and return back to a positive status once again.

"In other news, an arrest warrant has been issued for Alchemist and Mars Ayers. The space marshals have placed the Ayers brothers on space watch for conspiracy and numerous counts against everyone, including crimes of genocide, war crimes, and acts against humanity. In the case of Ayers Brothers versus the People of the State of the International Space Station, the International Criminal Court has already issued arrest warrants for fourteen individuals and summonses to twelve others. Five persons are currently in detention. Proceedings against one are ongoing, while four proceedings have been completed. Three have been convicted, and one has been acquitted while nine others are at large as fugitives. Mars and Alchemist were indicted during a pretrial chamber, where an arrest warrant was issued for the two brothers. Although the crimes that were committed were outside of the International Space Station, the International Criminal Court found that there are reasonable grounds to believe that these individuals had committed their crimes within the jurisdiction of the court. A three-billion-dollar reward has been placed for anyone who has any information on the two brothers.

"Next on Cyber News, have sex robots replaced or damaged human interaction between men and women? Our top experts will lead this discussion at the top of the hour. We are going to take a quick break. Stay tuned for more Cyber News.

***

Calvin had no luck searching for his wife and daughter. He asked for help at the police station and a couple of women's homeless shelters before checking into a homeless shelter for army veterans. Around ∇ 24:00, Calvin stood in the alley between the United World Station 13 Veterans Homeless Shelter and Big Joe's liquor store on Times Square Boulevard. The line was long, and Calvin didn't expect to get in that night; however, he was willing to wait it out. There were three staff members with clipboards standing in front of the line. They asked each person for a name and a photo ID. But before they let anyone in the shelter, the staff made sure that all children and women got in first. Once everything settled down, around ∇ 26:00, the staff let the last group of people inside to take the remainder of the beds. Luckily for Calvin, he was the last of the five men who entered the shelter.

"ID, please," said the staff member.

Calvin turned his forearm and showed the bar code of his

Benevolus identity card[8] with all his personal information.

"Calvin Davis," said the staff member. "Do you have a daughter named Skylar and a wife named Grace Davis?"

"Yes. Why?"

"Because they are registered in a room," said the staff member.

"Oh, really? Can I see them?"

"Sure. I can't let you in the west wing of the building where the women are, but I'll have another staff member inform them that you are here. We have a cafeteria room where you can meet up with your family," said the staff member. "But first, go inside, sit down in the hallway, and wait for further instructions."

"Yes, sir. Thank you."

"You're welcome."

---

[8] A Benevolus identity card is a microchip that is implanted in humans that is equipped with information about the individual that resides in the International Space Station and citizen of the United World. The information that is in the Benevolus identity card includes a person's identity such as name, sex, social security, and home address.

***

Meanwhile, chaos began to rise in The City of Neoterra after Apollo and the citizens had removed Alchemist from power. While Alchemist and his family were trying to find a way back to the International Space Station, his people were beginning to riot and destroy everything he'd built. Outside the palace, many people began destroying everything in sight. Whether it was statues, small building, obelisks, or large bodies of land, they burned it down while others charged inside Alchemist's palace. Once the bronze doors of Alchemist's palace were knocked off the hinges, Apollo led the chaos inside the palace. All their personal items were destroyed. Whether it was antique furniture, chandeliers and mirrors, they all were demolished and were broken. His self-portraits and family portraits were destroyed. His sculptures and reliefs were vandalized and defaced. All of Alchemist's classical oil paintings—Parmigianinos, Raphaels, Tintorettos, Leonardo da Vincis, Calvina Pozzos or photos of himself and his family—were destroyed. Portraits of his son, daughters, and wife were tossed into a pile of flames. Even Alchemist's pet lions, jaguars, tigers, cheetahs, and panthers were slaughtered. The people of The City of Neoterra had no mercy for

Alchemist and his possessions; neither did Apollo. Apollo strayed away from the group in search of Alchemist's bedroom. Once Apollo reached the bedroom, he stood in front of one of Alchemist's photos and began to speak as if he were looking in a mirror and saw his reflection.

"You coward. You were never the man who I thought you were. I modeled myself on everything you preached; however, in the end, your words were all lies. It was blasphemy for me to look at you as if you were a god, because you are just a man disguised as a demon. Now that you are gone, I will take over the city, and when I do, I will lead these people the right way. I will destroy everything you helped build and take over. I will start a new city, and I will leave my own mark. Mark my words. I will make changes that need to be made. I will keep all your scripts for all your speeches, every note and book that you ever created, because I do adore your work, and I respect your mind. However, you will be erased from the history of the city, and you will no longer exist to these people because I will replace your work with my name. I will carry on your genius work as a vessel; however, you will mean nothing to me or the city. Maybe I cannot erase what you have achieved from the mind of this generation; however, in generations to come, you will not enter the minds of my people. It's over for you, Alchemist! You are dead to the city, and most importantly, you are dead to me!"

Apollo reached for the portrait from the wall, grabbed his pocketknife, stabbed the photo, slit it open, and then tossed the photo through the window into the pit of flames outside the palace.

***

The chaos heightened when the group of rioters started to gather women who were allegedly pregnant with Alchemist's children. Prior to Alchemist being forced into banishment, a lot of women in the city tried to get impregnated by him. It was an honor because of how much power Alchemist had at that point. However, since he had been banned from the city, it was now frowned upon. A lot of the women who were being rounded up were being turned in by other women who they knew and had hated them, so they had no problem getting them involved in all the chaos. One of the women was a young lady named Rose. Rose was one of Alchemist's concubines. She was only eighteen years old, and she was six months pregnant with Alchemist's son.

"If you allow us to abort your child, we will spare your life. All we need is the child. That's it! That's all we need," said rioter number one.

"But I'm pregnant."

"Don't worry. We will provide you with the best doctor in the city to perform the procedure, so you will have nothing to fear."

"No! No matter what crime Alchemist has committed, this is still my child, and I will not put my child in harm's way because of what his father did. If you want to discipline anyone for his actions, find Alchemist and punish him—not my child."

"We understand that is your child. I understand because I have children of my own. However, there is no reason to protect your child because his blood isn't pure! He is the son of Alchemist!" shouted rioter number two.

"Like Apollo said, we must protect our generation to come. What if he grows up and finds out that his father is Alchemist? Like any son, he will try to avenge his father's name if you raise that child. We understand what you are going through; however, we must protect the future generation," added rioter number three.

"No! I don't care. I can't live with myself knowing that I gave my son's life away because of what his father has done. I am sorry! If you take my son's life away, then you have to take my life as well."

"Enough said. Take her away and tell the others to prepare for the death of Alchemist's son!" said rioter number one.

"That's fine—I wouldn't ask to die any other way. I'd rather die with honor than live in disgrace."

The rioters carried Rose to the center of Alchemist's courtyard right next to the pit of flames and placed her on the ground where they began to gash her open. Once they cut open Rose's womb, blood splashed everywhere as they tried to pull out Alchemist's child. Once the child was visible, they pulled the infant from the wound and placed him on the ground. Then everyone began stabbing and stomping the infant as Rose bled to death. Once they finished with Rose, both bodies were dumped in the pit of flames. The rioters continued with their madness as the bodies of Rose and her son turned into ashes.

***

As the chaos and the mayhem slowly died down, an aircraft appeared in the sky. No one was sure what it was until everyone approached the object that had crashed miles away from all the madness. It was a space capsule with a family from Planet Nibiru inside. The Mason family was one of many families leaving the International Space Station looking for a new planet to live on because of the dwindling food supplies and environmental crisis. There was a diminishing amount of water, food, oxygen, and other natural resources in space that caused everyone to flee the International Space Station and take refuge on Planet Nibiru. However, once the Nibirians[9] figured out the agenda of the humans, they began creating changes that would affect humans, and the family left to find refuge on Planet Earth.

They had never been to Earth before, so everything was new to them. As the family slowly stepped out of the aircraft, they were intrigued by their new surroundings. Unlike the atmosphere in the space station, everything was organic. Whether it was the grass, the mountains, or the air, it was immaculate to the family from space.

---

[9] Natives of Planet Nibiru.

For a moment, the family members were in a daze.

However, once they were able to gather themselves, they began to greet everyone surrounding the space capsule. Apollo had seen the aircraft land, and since he was the leader, he felt obliged to introduce himself as the leader of the city. Once Apollo introduced himself, the Mason family introduced themselves to Apollo.

"Greetings. My name is Sebastian. This is my wife, Ayda; my daughter, Xyla; and my son, Zahi. We are the Mason family. We are looking for refuge from the crisis on Planet Nibiru."

"Well, you came to the right place. Hello. My name is Apollo; welcome to my city. Here, come with me. I have a place for you and your family to settle down..."

CHAPTER TWO

Episode 11: Agenda

In a life span of a lion, the lion will show two sides of itself. The first side is the nurturer, and the second side is the beast. The nature of raising a pride of lions is to raise the cubs by nurturing them with care by feeding and protecting them. However, for a lion to protect and feed its pride, the beast must come out along with the nurturer.

Audio Entry #05

"At this moment, everything seems good with the family except Scarlett. Back when I and the family were living in The City of Neoterra, Scarlett was faithful to me, even when I was wrong; she always had my back. However, now I don't know if she feels the same way she did in the past because ever since we lost all our fortune and wealth, her attention has changed. Scarlett was losing faith in me, and she was questioning my greatness.

I notice that now that I have no power or wealth, she treats me differently. I feel like I am losing Scarlett's trust and faith in me and that she'd betray me if she had the chance. The worry I saw in her eyes when I tried to help Sade concerned me. Maybe I am just paranoid; hopefully I am wrong. However, only time can tell."

Back on Planet Earth, as Alchemist and his family returned to their Space Arc, two Lotus' were scavenging in the forest. As one was in search of its next prey, another was about to meet its fate. The first Lotus was the alpha female. She was feared by all species and breeds that inhabited the mountain. She was like no other creature on the mountain. The red Lotus was one of a kind. She stood tall, and her physique matched her bravado. Her thorns were as sharp as her mind. Her skin was as red as blood. Her boldness was as broad as her shoulders and her chest. Her confidence was as extraordinary as her abdominals, which were a sight to see, but not as great as her sculpted thighs.

As for the other Lotus, she was barely half of what the alpha female Lotus was. The other Lotus wasn't as developed as the red Lotus. She was only two years old, and his mind was still innocent, playful, cheerful, and joyful. He was not battle tested, and his killer instinct was not developed yet. His thorns hadn't been fully developed yet, and his scales were smooth. His scales were powder white, which made for an easy target.

The red Lotus found thrill in an easy kill; from a distance, the red Lotus prowled toward the young white Lotus as he picked through a bush of blood oranges. As the leucistic skinned Lotus chewed her orange, he stumbled and then gathers himself. It was charming; however, the sight of weakness excited the red Lotus and forced him to attack. The alpha male predator jumped on top of the weak and ripped her into pieces. He ripped her soul from her body, spared her flesh for her cubs, and left her skin and bones to scatter

for the vultures.

***

A few miles away, Alchemist and his family continued their quest to space. Soon after the sun had set, Alchemist and his family left Abe Smith's farmland and began traveling toward the Space Arc. At dawn, they were caught in the center of the snowstorm, but they were able to pace themselves through the mountain and reach the Space Arc. As soon as they arrived at the crash site, his family began settling in as Alchemist began to rebuild the Space Arc.

"Finally, we have arrived. The spacecraft is in worse condition than we left it; however, there is no reason to be discouraged. It's been a long journey; however, we should rejoice. But there is no time to rest because we all have a task at hand. We are close to reaching our next destination, which is space. However, we must work together and put in the work to get where we need to be. I am going to work on rebuilding the spacecraft; in the meantime, I need your help to gather a few things. I am going to assign you each a task. Carmen, I need you to find a river or a brook to collect some water. Hunter, I need you to help your mother in gathering some food and animal fur for the entire family before sunset. If you need some supplies, check the supply room in the spacecraft. Let's go and do this now."

As Alchemist and his family began to settle down in the spacecraft, they did not realize they were about to walk into a rude awakening. The family entered the spacecraft and went to their cabins. But once Alchemist entered the cockpit, he realized that Lotus' inhabited the spacecraft. The room was filled with the aroma of defecation, urine, and amniotic fluids. As Alchemist looked under the control panel with Sade in his arm, he discovered six baby Lotus' lying on a pile of sheets and clothes in the center of the room. At first, he didn't know what to do. Alchemist first guarded Sade from entering the room, but once he realized that the room was secure, and the babies were harmless, he let his guard down. Alchemist let Sade hold the babies. Sade cradled a baby in her arms and began petting its smooth scales as it gurgled and purred, but soon she caught her hand in its premature thorns. As soon as this happened, Alchemist took Sade away from the Lotus' and found an empty basket where he laid the babies to rest with covers and blankets. Later, Alchemist was able to generate power in the Space Arc and reconnect the AERO interface; the backup generator turned on and

disabled the power on the left side of the spacecraft,

so no one could be electrocuted. The engine was running with the

coolant circulating to generate heat in the interior of the vehicle.

Later, Alchemist settled Sade down in her own cabin, where she had a vision of when her father was a child. In her vision, during a Montessori lesson, Alchemist was sitting down working on a ditto to solve the famous mathematical problem of Fermat's last theorem, and he was not allowed to leave the room until he solved the problem. Young Alchemist began at six in the morning. Now it was seven o'clock in the evening, and he was still working on this problem as a punishment. The teacher was trying to discipline him; however, Alchemist's mother felt another way would be better. Alchemist's mother had sympathy for him; she felt his teacher was being too rough. So as Alchemist tried to solve the problem that was in front of him, his mother came into the room to let him go to his bedroom.

"You're wrong again! I'm telling you that you won't leave this place until you get this problem right. Get it right! Do it the way I taught you!" said Alchemist's teacher.

"What's going on? It's seven o'clock. Alchemist, what are you still doing here? Go to bed, now!" said Alchemist's mother, Bella.

"I'm sorry, Mrs. Ayers, but I can't let you do that; he must solve this problem before he can go and rest."

"I am his mother, and I say that's enough."

"But his father won't be pleased. He assigned me to teach Alchemist this lesson, and he was adamant that he does not leave until he solved these problems."

"That's OK; I'll deal with his father. Go ahead, Alchemist; go rest."

"Yes, Mama. Thank you, Mama."

However, when Alchemist's father learned what his wife did, he wasn't happy about it. Alchemist was going to participate in the big International Space Station mathematics competition, and he wanted him to be prepared to win. Zeno had high expectations for his son and was putting a lot of pressure on him. However, Bella didn't feel he was being fair and was trying to reduce the pressure on her son. But when he found out, he scolded her in front of Alchemist.

"Where is Alchemist? Why isn't he in that room learning what I assigned Ben to teach him?" asked Zeno.

"He's resting. Ben was working him too hard, so I sent him to bed," said Bella.

"Are you serious? Alchemist should be practicing mathematics," said Zeno.

"Well, I wanted him to rest. Besides, he'll be fine. I believe in him," said Bella.

"Well, I'm not going to take the risk. Alchemist come here now!" said Zeno. "We're not going to do it your way. We're going to do it the right way, which is the Ayers way. When I was Alchemist's age, my father did the same thing for me. Every competition, I would practice until I couldn't think anymore. I ate and shit everything I did, and that's what I'm going to teach my son. He's not going to think quitting is OK because his mother said it's OK. He's an Ayers, and I am going to raise him as an Ayers."

"Yes, Dad."

"Go back and finish that ditto, and if your mother says otherwise, ignore her. Listen, Bella, I respect where your heart is. However, Alchemist is carrying our name on his back; we can't put him out there knowing that our child is not ready. We must prepare him for the real world before it's too late. Got it?"

"Yeah."

"Good. Don't you ever question my authority again."

***

Soon after his speech, Secretary-General Zeno remained in the Peace Palace to get some feedback. Reaction to the debate was positive. However, everyone's focus wasn't the speech. Since the indictment of Mars and Alchemist, everyone had been talking about whether Zeno was an accomplice in the crimes that his sons were being accused of. Zeno tried to eavesdrop and figure out what everyone was talking about.

"What did you feel about the speech?"

"If I was his publicist, I would have told him good job in sticking to the script. He read his speech and then answered all the questions without breaking a sweat. However, that's not going to stop people from talking."

"I know, hypocrite; how can he change the issue that is going on in the universe when he can't control what is going on in his own home?"

Every time he walked by, they refrained from talking or simply walked away. It wasn't until the second-in-command, James Aesop, pulled him to the side and explained what everyone in the

Peace Palace was talking about that he knew.

"What's up, James? What's going on? What's everyone talking about?" asked Zeno.

"To be honest, everyone is talking about you and your two sons. Everyone is wondering whether you took part in this plot that plagued that city on Earth. Everyone thinks you used your sons to do your dirty work," said James Aesop.

"Why would they think that I had something to do with what happened on Earth? I had no idea about what they are being accused of doing."

"You have to understand that there is a conflict of interest: your sons are accused. Plus, you're the only one who has access to all those weapons. I personally don't think you had anything to do with it, and I can vouch for you. However, we must prove your innocence.

Have you heard from them yet?" asked Aesop.

"No, I haven't."

"That's not good, Zeno. I must warn you. If you don't bring Alchemist and Mars into custody, first you will be forced to step down, and then you will be indicted for conspiring to harbor a fugitive and to lead a genocide enterprise because you are Alchemist and Mars's father. I know you had nothing to do with what happened, but it's not up to me; you must prove your innocence. You have to prove it to them," said James Aesop.

***

Mars and Pluto finished their meeting, and as he walked down the grand stairs, Mars noticed the grand piano in the room.

"By the way, is that a Steinway Alma Tadema? Do you mind if I play a few songs?"

"Sure, knock yourself out."

Mars played the beginning of the piano version of Bruckner's Symphony no. 8 in C Minor (*Apokalpyptische*) before he went on his way to Planet Earth.

Finally, Mars arrived on Planet Earth, courtesy of Pluto. His mission was to find and then kill Alchemist. Pluto ordered him to stay put until he was given further instructions; however, before he tried to find his brother, Mars decided to take a detour to visit his mother's grave. Mars landed a few miles outside the New Apollo; although his mother was buried two hundred miles away, that didn't stop him. After he looked at the city, he reentered his aircraft to travel to his mother's grave.

"In other news, Planet Europa has struck a deal with the Republic of Pluto, in which the planet will be supplying the interstellar nation with a pipeline of water. This deal is estimated to be worth over two-point-two quadrillion dollars. We will keep you posted with any further developments in this massive deal."

As the news about the deal between Planet Europa and the Republic of Pluto broke, Angel was intrigued.

***

(Knock on the door.)

"Who is it?"

"It's me, Adakis. Father, may I speak with you?"

"What's on your mind, my son?"

"What was that all about with you and Mars?"

"Nothing that concerns you."

"Is it about Alchemist? This is family business. Why are you getting an outsider involved when I can do whatever you're telling him to do?"

"Listen, son, right now you are not able to ask any questions. You went behind my back and did things that I never told you to do, and now you have the audacity to question my authority. Who the hell do you think you are? You follow my orders, and you do what I say—that's it. See, this is the reason you were demoted from the army. You don't follow orders."

"I do listen, but you never hear me out."

"Do you remember what you did? You were suspended from the space force because you brought Angel to me, and you convinced me to bring him in. Then you allow him to give orders to my army without my approval. You deceive my sergeant in command by lying

to him, telling him that I had given those orders to him to send a

group of my men to Planet Europa and attack our prime source of

water, steal it, and bring it back here. And you still haven't found

Angel. Where's Angel?"

"I don't know."

"You don't understand how detrimental this was to us as a

nation. Because of what you did, this started a conflict between us

and Europa. They were once our primary source of water, but

because of you, Tabbris cut all pipelines to us. Do you understand

how serious this is? You could have started a war between us and

Europa. This type of war will never end because blood is involved.

Luckily enough, I was able to be diplomatic and covered your ass by

resolving the issue. I spoke with King Tabbris. However, we still

don't do any business with them because they don't trust us anymore.

Now we have to find another pipeline for water before we run out."

He sighed. "I swear, you're lucky that you are my only son,

because if you weren't, I would have killed you a long time ago. You

brought shame to not only our family but the republic as well. I don't know if I can ever forgive you for that. Is there anything else you want to talk about? Because there is nothing else to discuss."

"No, sir."

"All right; you can leave now. I have things to do. Make sure when you leave that you lock the door behind you."

After Adakis confronted his father, he walked out of the room and tried to turn the corner, where he bumped into his mother, Alexandria, who was eavesdropping on the conversation, trying to figure out what was going on with Mars and Pluto.

"So, what happened? Did he tell you what that was all about with him and Mars?"

"No, he wouldn't tell me anything. When I tried to figure out what was going on, he brought up the incident that happened between me and the sergeant."

"Interesting. I knew he was going to do that to you, but don't worry, I know what to do. You'll get your chance to prove to your father that this is your empire and you are worthy of being a leader. If that means we must go behind his back to do it, then we'll do so. In the meantime, just stay put. Once I get the information, I will tell you what to do next."

In the distance, Angel was listening to Adakis's conversation with his mother before approaching him.

"I can't stand your father; he is such a brute. He is filled with nothing but testosterone. Reminds me of my father. So, what are you going to do?" asked Angel.

"I don't know yet. We're going to wait and see."

***

Back at Capital Hall, after landing in The City of Neoterra, the Mason family was brought to the east wing of the executive mansion, which was the least damaged part of the building. The Mason family was given a hostess to help guide them through the madness and explain everything that was going on in the city. As the Mason family walked through the executive palace, Sebastian had a few questions for the hostess, and with the help of the hostess, they learned more about Apollo and his campaign to become the leader of the people.

"Hello, my name is Sierra. Welcome to Earth and The City of Neoterra. I will be your hostess, and I will be giving you a tour around this beautiful city."

"How long have you've been living here on Earth?"

"I've been living here my entire life."

"Wow, I can't believe it. Everything Alchemist spoke about in his speech was true. People actually do live on Earth."

"What speech are you talking about?"

"Alchemist gave a lecture in which he tried to prove that

Planet Earth was inhabitable again. He sent it to everyone in the universe and tried to convince everyone to return to Earth, but no one believed him. They're lying to us about Planet Earth; there are regular people like us living right here on Earth, but for some reason they didn't want us to come here. I guess they were hiding something from us or hiding something from the people on Earth. I think we had to see it to believe it."

"But what happened in the past for people to think Earth was uninhabitable?"

"Well, you're probably too young to know anything about the nuclear war in 2050, but this is what caused the planet to become uninhabitable. So much damage was done that everyone had to move to outer space. That's when most of humans moved to space."

"But not everyone made it to space. When the people left to go to space, they cut all our resources and left us stranded to perish on Earth. However, as you can see, we persevered. We were able to get back on our feet all because of Alchemist."

"You know Alchemist?"

"Yes, he was our leader at the time. Alchemist was a very important figure to us because, after the apocalypse in 2050, I learned that a lot of government officials and other important individuals fled Earth and left it in a shamble. Before Alchemist arrived from Earth, the people who were living on Earth were living in primitive conditions. However, when Alchemist came, he rebuilt this area into a fully developed industrial city with a capitalist system. In the beginning of The City of Neoterra, with the help of Alchemist and his knowledge, the city became dependent on the production of commodities through wage labor in the interests of capital accumulation.

"In only a couple of years on Earth, Alchemist was able to first rebuild and improve upon the ecosystem and then create his own macrosystem. He created a division of labor for different trades and the principles for each position. This method helped increase production and create work for the citizens in the city. Alchemist developed a source of animal and human labor and manufacturing productivity using energy from fossil fuels, atomic power, and solar power to foster economic growth. Alchemist was able to organize three different social classes based on the people's skills. Those who were skilled in agriculture or manufacturing or any other physical labor were a part of the labor class. Those with professional skills, such as teaching, trading, or municipal or public affairs (politics or military), were a part of the middle or wealthy class. Alchemist structured the society on classical free-market fundamentals.

"Next, Alchemist created a currency to serve as a common medium of exchange. This is what helped individuals to be able to work in any position and make any exchange without barriers. The City of Neoterra's urban cultural role fit perfectly with the capitalist economic order that came to dominate all other social institutions. With his knowledge, Alchemist was able to provide a health system, a court of law and code system, an organization of urban citizenry forming a sense of municipal corporations, and a Montessori school system to teach children math, science, English or world language, and history. A chapel was built to teach religion and moral standards. There was adequate political autonomy for citizens to select the city's governors. There was an army and police force and a market for the exchange of goods."

"You know, what people in space say about people on Earth isn't true. You are not primitive after all. Wow, I can't believe that Alchemist really lived here on Earth and was the leader of a city. Some people thought he died on his way here; others thought he never arrived on Earth. I guess he wanted to follow in his father's footsteps and build his own nation. But if Alchemist did all these great things, why are so many people angry and rioting?"

"It's because he betrayed us and lied to us. That's why Apollo is our leader now."

"And who is Apollo? Is he that young gentleman who introduced himself?"

"Yes. That was Professor Apollo."

"How old is he? He looks mighty young."

"He's twenty-four."

"Wow. How can one man so young have so much power? It's like he's a god in the flesh. The people look so angry, but he seems so much in control."

"Yes, indeed, he once was one of the prime professors in the city. He led us to victory and helped us break away from the Alchemist regime. When the Alchemist regime was in power in The City of Neoterra, Apollo was a highly respected individual. Apollo earned the respect of the city for what he was able to achieve while Alchemist was still in power. He was one of the preeminent scholars in The City of Neoterra. He was chosen by Alchemist to be a part of his elite group of scholars who would be able to follow and learn

from Alchemist himself. Apollo was one of ten individuals who had

the privilege of following and learning from Alchemist as a candidate

to be a professor for the Alchemist Academy, and the young Apollo

had the potential to become the next leader of the entire academy.

He lived in the shadow of Alchemist; every day Apollo strove to

practice everything Alchemist preached and thought. Apollo

respected Alchemist. From afar, Apollo followed Alchemist.

"When Alchemist was a Montessori teacher, Apollo was

present at every lesson and lecture. When Alchemist was a preacher,

Apollo was present in every congregation. When Alchemist needed to

make a speech to the people, Apollo was there listening. Alchemist

was once someone he modeled himself after; however, now he has

become everything Apollo dislikes. Apollo began looking at

Alchemist as someone who was materialistic, shallow, promiscuous,

greedy, thieving, and prideful. Once his views on Alchemist changed,

his attitude toward life changed. Apollo was a quiet, gentle, and

submissive individual; however, once he lost his innocence, his

motive in life changed. He became violent, with uncontrollable anger.

He began to resist authority, control, and convention. He became

ferociously aggressive. His goal in life became to destroy everything that represented Alchemist and everything he helped build. Once he destroyed everything that Alchemist represented, he decided he would take over Alchemist's city. Apollo was ready to face and endure the danger that would be ahead of him once he took Alchemist's position, and now he is going to make things better for us in the city."

"And who is that fine young lady standing by his side? Is that his significant other?"

"That is Zyana. I believe she will be crowned by Apollo to be queen of the city very soon. Zyana and Apollo have been together since they were young. Apollo always was attracted to her piercing green eyes, high cheekbones, thick lips, and slim figure. Who would resist making that beautiful woman their queen? We'll see if he crowns her; only time will tell. So where are you from?"

"We are from the International Space Station. We were born and raised as United World citizens. However, after the living conditions began to deteriorate, we ventured to Planet X as the new planet host."

"Are you humans?"

"Yes, we are humans."

"OK. Where is Planet Nibiru?"

"Four hundred light-years away from human existence."

"Oh wow. What type of planet is it?"

"It's a spiritual world where a forbidden land rests. This planet is called Nibiru (referred to by humans as Planet X), and on this planet, there is a forsaken land called the city of Utopia. Planet Nibiru is thirty-four light-years in length, and within this planet there are thirty-four goddesses—servants of the Alpha god. The goddesses each control a different physical element of all planets in the universe: earth agriculture, gems, and jewelry, water, fire, air, space, darkness, and ether (the spiritual component of the soul). The Nibirian goddesses are the supreme guardians of light, time, and power structures and the keepers of reason, truth, wisdom, and purity. When the gods come together, they form as one body. However, Queen Nebula is the only goddesses who can bring them all together and exploit all the elements and the powers of the gods.

Queen Nebula is the reincarnation of the gods of Planet Nibiru and Beta of the universe. She was made from the immaculate conception of Alpha god, and her job is to create peace in the universe.

"Queen Nebula carries and controls the souls and energy of all thirty-four goddesses. Queen Nebula has the power to guard and regulate the stars, the seasons, and the nature of the entire universe. And she is the only goddesses who can appear before other lifeforms. Queen Nebula can use her power and appear among other life-forms by choosing a Nibirian person, a titan, amazon, or even a human and giving him or her divinity from the gods. She can also arise in a physical form as she chooses but only to defend or establish balance and order in the universe. Queen Nebula is the caretaker of beauty, love, and healing. However, Queen Nebula also mirrors the strength of the gods as well. Queen Nebula has the willpower to bring the darkness out of all elements and use elements as weapons. When needed, Queen Nebula can be fierce and can exhibit unmatched wrath. The only time she shows her strength is in times of war.

"Planet Nibiru and the city of Utopia are places where everyone is free from all mental and physical suffering, and in this place, Nibirian goddesses, Nibirian amazons, and the souls of the dead live and worship among one another. City of Utopia is also a place where the secret powers of the gods are guarded. There are two bridges between the two worlds: one is spiritual, and the other is through the gateway. If you were not born in this world, for you to have access to the city of Utopia, you must be heaven sent. Queen Nebula holds the key to the gates of the City of Utopia. Demons and mortals are forbidden access to the city of Utopia and Planet Nibiru. However, one of the ten sages of Planet Nibiru has prophesied that demons will return to the city of Utopia and destroy everything the gods have built. This will start a holy war among the demons, the mortals, the Nibirians, and the gods. However, because this story has been passed down from generation to generation, they know what's coming and are prepared for the worst."

"And you traveled all the way from Planet Nibiru? How did you survive that travel without dying?"

"The reason we were able to survive the distance we traveled is because our spacecraft was built the same way the International Space Station is built. But it sucked living in space. I have only been here for a couple of days, and I feel like I never felt before. It's wonderful here on Earth; as soon as I stepped on Planet Earth, I could feel that we caught the old wind feeling of being here. Everything is organic; when we stepped out of our spacecraft, I had a feeling of pure ecstasy rushing through my body from the feeling we got from the air. On our spacecraft, we did things like inhaling flavored oxygen or refilling our bodies with Myers's cocktail because we had been deprived of these things back in space. Things like eating were a means of survival, but with the rich fresh food at our disposal now, we've finally found joy in eating. I found an apple, and when I ate it, it was like I had an orgasm on my tongue from the fruit."

"Is the International Space Station in space?"

"Yes."

"How does one live and survive in space?"

"Everything is synthetic: from the things we eat to the surface area of the space station to our atmosphere!"

"And where is it located?"

"Right next to the moon. The International Space Station was once an artificial satellite. However, it's now an interstellar ark. I don't know if you've read the Bible…"

"Yes, I have."

"Well, the International Space Station was like Noah's Ark, created for the same purpose. It was predicted that biological warfare would ensue among the late United States and other power nations.

So, a group of scientists, engineers, and architects decided to come together, and they were able to expand what already existed in the International Space Station and create sort of a spacecraft city/microcosm, built with an ecosystem that was like the chemical elements of Earth to ensure the survival of humanity in the event of nuclear war—if Earth became uninhabitable. The chemical elements include oxygen, silicon, aluminum, iron, magnesium, calcium, sodium, and potassium. The International Space Station is also equipped with a fully developed ecosystem that can generate light, air, water, food, and gravity. With aero terrestrial soil, we can grow organic matter such as biofuel and sustainable food."

"How long has this International Space Station existed?"

"Close to a millennium."

"Wow; that is amazing. But how is this space station able to remain in space for so long without running out oxygen or fuel?"

"The International Space Station is equipped with a power plant that's four thousand meters in diameter with millions of solar panels that help generate energy and give electrical power to the space station. To fuel the space station, asteroid mining is used for raw materials—water, oxygen, and many other natural resources."

"Wow; the International Space Station sounds incredible. But why did you decide to leave and come to Earth?"

"The reason we are here now is that conditions in space have changed for us. The space station is in decay, and people can't live in space anymore; the population has increased, and there is a shortage of resources. That's why we went to Planet Nibiru; we were looking for a new planet to live in. Rich people emigrated to Planet Nibiru and built a new empire while the poor remained in the International Space Station. The ones who were able to move to Planet Nibiru began to build a relationship with the extraterrestrials.

"Wow. That's amazing. How was life in Planet Nibiru?"

"At first it was great, but once the humans had betrayed the Nibirian people, there was a war that broke out between the two races, and the humans lost. The humans were enslaved, and the Nibirian people forced people like us, who were just coming for a better living, to work for free to help pay back all the damage the humans caused when they were on Planet X. Once we realized that we lost the battle and the natives began capturing human explorers and making them indentured servants, we knew we had to leave. However, by the time things got serious, it was too late. The entire human race on Planet Nibiru had become slaves to the Nibirian people.

"Were your family members slaves?"

"Yes. Humans were treated harshly and with no dignity. The condition for humans was very difficult. I couldn't witness my family being mistreated—they were beating us, raping us, and murdering us on Planet Nibiru—so I decided to leave. We managed to escape Planet Nibiru; however, we had nowhere else to go. We could have returned to the International Space Station, but we had nothing there. We had already gotten rid of all our belongings, our home, and our vehicle; we sold it all, so we could migrate to Planet Nibiru because we were fully invested in moving back. There was no other planet, so we took a risk and decided to come here to Planet Earth. Legend told us that Earth was an uninhabitable planet that wasn't safe for humans to live in. However, we felt we had no choice but to try Planet Earth as our next host." "Well, you made the right choice."

Sierra showed the Mason family a view. "The city is divided into three zones, and in each zone, lives a different social class. Over there is zone three. This is where the lower-class citizens of the city live. If you look over there, that is zone two, the industrial section.

This is where everyone works and some live. And over there is zone

one. That is where Capital Hall and the palace are located. That is

also where Apollo will be giving his speech tomorrow."

*****

Back at Station 13, the capital of the International Space Station, when Calvin entered the shelter, everyone's personal belongings were inspected. Once everyone went through that process, they then entered the homeless shelter.

Once everyone was settled, the staff gave Calvin and his group a step-by-step rundown on how things worked at the homeless shelter. "All right; you guys can sit over here, and someone will be with you in a short moment."

"Hello, you guys. My name is John. First things first: I have some papers that I need you all to fill out. This paper right here is an emergency-shelter admissions form. After you sign this sheet and agree to our terms, we will place a signed copy in our files. In the meantime, while you get situated, one of our staff members will call you in to talk to the shelter director, who will ask about your current situation and how we can assist you. We will review your plans and goals. Everything will be confidential. Make sure you have two forms of identification. One must be a photo ID. Once you are approved by the shelter director, you will be allowed to stay here at Station 13's veterans' homeless shelter.

"This visit is temporary until you find a permanent home. You're here for a short-term stay, so use it wisely. You have only ninety days at this shelter. An individual is allowed no more than three stays in a two-year period. There must be at least four months between stays. Families with children may be approved for extended stays of one hundred twenty days. If you are unemployed, you are required to seek employment and permanent housing while staying here and turn in a job sheet and a housing sheet. If you don't submit any of the sheets, you will be withdrawn from the program, so make sure you do that. Check in with the shelter director or executive director when reporting for work. The doors will open at five thirty. Lights go out at ten o'clock. Lights come on at six in the morning. Residents must leave no later than seven every day. Everyone here will be given the opportunity to shower every day around seven in the evening. Make sure nothing wet or dirty is left in the bedroom or bathroom area; make sure you keep everything neat and clean before you leave the building. We provide clothing, towels, and personal hygiene products, if needed. If you need to use laundry facilities, you must be finished prior to ten o'clock. Food and drink are not allowed in the dorms. Our staff will provide breakfast in the morning and dinner at the soup kitchen. Any questions? All right; now for the rules of the shelter.

"No smoking, drugs, or alcohol in the shelter. If we suspect you're using, you will have to perform a Breathalyzer or drug test. If you fail, you will be asked to leave the shelter. If you have prescriptions or are taking any over-the-counter medications, it is your responsibility to submit all those items to the nurse on the second floor.

"No weapons of any kind are allowed. If we find one, you will be asked to leave. The shelter is not responsible for personal items. Personal belongings that are left for twenty-four hours after you leave the shelter will be thrown away. Everyone who enters the shelter is permitted to have only two large traveling bags and two duffel bags. You are also allowed to enter with personal items in two trash bags.

"You are not allowed to use the phones longer than thirty minutes or to make any long-distance calls. If you do, you will lose phone privileges.

"And last, no sexual activity. As a government-funded organization, we are required to provide for you as citizens of the United World; however, if you don't follow the rules, we have the right to not admit or to remove you. Any violation of any rule may result in disciplinary action up to and including dismissal from the shelter. People dismissed for violation of any rule may not be able to return to this shelter. Any questions? All right. In the meantime, once everyone is processed, I will be giving a tour of the facility, and then you will be assigned to a dormitory. Follow me," said John.

***

As Dallas traveled toward Alchemist's secret lab, he watched the latest news report about his family. Michelle Waters was reporting new information on the massacre case.

"Toxic weapons and weapons of mass destruction—these are the tools that were used in the infamous genocide a year ago. There is new information about the Metropolis City massacre. Authorities say that there is a possibility that Secretary-General Zeno Ayers was involved in the mass murder that took the lives of thousands of innocent individuals on Planet Earth.

"Mars and Alchemist have been placed on the United World space watch for conspiracy in leading a democide group. Evidence shows that highly lethal weapons were used during this mass murder, and questions remain about how these two individuals gained access to these weapons of mass destruction. Our source believes that Secretary-General Zeno Ayers had approved his two sons' attempt to take over Planet Earth and use it as a host. As the International Space Station begins to become uninhabitable, sources believe he is planning to wipe away the existing humans to relocate citizens of the

United World to Planet Earth. We will continue covering this story as new information comes in."

After watching what the news had to say about the situation, Dallas decided to make a detour to the Cyber News headquarters to clear his family name. While the Cyber News was filming live, Dallas rushed into the studio for an interview and pounded on the door. "Let me in! I want to speak to whoever's in charge! I saw what you guys were saying about us on the news!"

"OK, OK, calm down.

At first, the producer hesitated to give him the opportunity to give his side of the story. However, after he thought about it, he would give it a chance.

"We'll give you an opportunity to tell your side of the story."

Dallas could get the interview he wanted. They cut to commercial, and the producer equipped him with microphones and earpieces.

"OK, so this is how it's going to go," the producer instructed him. "You go in there and sit down in the chair. The reporter will ask you a series of questions, and you'll have a certain amount of time to answer, and then boom, interview's done. Just make sure you don't use any kind of profanity. Got that?"

"Got it."

"All right then. Let's do it.

***

After Queen Nebula was banished, Iris took full control of
Planet Nibiru. The first thing Iris did was enter the forbidden temple.
Trespassing in the city of Utopia was strictly forbidden to those who
were subject to death; only the gods could enter. However, now that
Queen Nebula was gone, Iris had access to the temple. In the
forbidden temple, it was said that the spirits of the gods lived. It held
the power of the gods. The chosen one would be given forsaken
knowledge and power from the Alpha god and would be able to
control all elements of life: all physical elements (earth, water, fire, air,
and ether), all senses (smell, taste, sight, touch, sound), locus of
control (power, knowledge, desire, space, time, life/death, and
good/evil), and the soul (energy and intuition) if they entered the
temple. With the help of his Head Seraph, Sirus by his side, he
entered the temple. In the process of entering the forbidden temple,
Iris opened a vault where the book of Utopia lay. There were
duplicates of the book of Utopia; however, they were all missing an
important verse. He immediately walked toward the book of Utopia,
opened it, and then began to skim through the entire book. As Iris
was skimming through the book, a passage from the book caught his

eyes. Throughout his life Iris was taught that the Nibirians had reached the highest point as a creature and were supreme beings. It was passed along from his ancestors, and he believed and lived by what he was taught. Iris opened the book. The script revealed that all living creatures were born mortal on Earth, where they lived as beasts.

Iris shut the book closed. He couldn't believe that he evolved from a beast.

"I am a Titan. I am a descendant of the gods. There is no way I came from a seed of a beast."

But he wanted to know more, so he reopened the book and finished the passage.

After death, they are resurrected, and they evolve into human beings. Then their spirits are sent into the spirit world in their human afterlife where they are judged by Amazons and Titans. Once Amazons and Titans allow them to pass the pearly gates, they are given a new name and become Amazons or Titans themselves before turning into Nibiru stars. Those who live in the spiritual world will be blinded by the craftiness of men, and their main purpose is to rejoice and worship god and give guidance to others who want to reach the promised land. Those who were sent to in the spiritual world will remain until they are judged at the final judgment where they are chosen to become Nibiru stars for eternity. The golden star is the source of energy and the soul of the Amazons and Titans. Without the golden star, the Amazons would not exist. Like the sun is to humans, it is the golden star that gives light and life to all Amazon and Titans, Nibiru Stars, and the rest of the universe. Those who have been righteous during their lives or in the post mortal spirit world become Nibiru stars and become the source of individual energy for the golden star.

However, those who sin against the Alpha god will return into the crust of the earth where the individuals who commit unforgivable sins will be sent into a spirit prison for a thousand years and will be servants of God. If the individuals redeem themselves, they're sent back to the surface of earth where they begin the process as mortals all over again. If not, they will remain another thousand years before they are judged once again.

Iris closed the book once again. Once he found the sacred box where the golden star lay, he opened it. The golden star is a sacred gift from the Alpha god and only the chosen one would receive revelations and power from the Alpha god. At that point, Iris knew the truth about promised land. However, instead of telling the people, he created his own script and posed as the Alpha god. Then he stole the power from the golden star before he took away all powers and dethroned the goddess.

"With this golden star, I will be held as Alpha of Planet Nibiru, and Alpha of the universe. You all will feel my wrath and worship the ground I stand on."

Now that Iris was Alpha of the universe, he was invincible, and no one could stop him.

***

Audio Entry #06

"My main mission is to keep the family well clothed and fed. But in the meantime, I am going to take this opportunity of being out in the wilderness, and with the help of Scarlett, I want to teach my son how to hunt. To avoid overkills, instead of using a laser gun, Scarlett is going to teach Hunter how to hunt by using a bow-and-arrow laser that was stored in the Space Arc. I pray everything goes well, and he comes back in one piece. I know Scarlett will."

Once Scarlett was able to gather all the attachments for the bow and arrow, she and Hunter were up and hunting. At the break of the dawn, Scarlett and Hunter were out in the wilderness. At dawn, they returned to the Space Arc. With a single laser bow and arrow, Scarlett was able to hunt down a couple of mule deer, beavers, skunks, wolves, and brown bears. From a bald eagle's nest, they were able to collect a couple of eggs. From a mountain stream, she caught fish and collected water. Minutes after coming back from hunting, once she examined the meat for any infections, worms, parasites, and diseases and cut the musk glands, she cooked it and served to her family. From the skins of each animal, she created garments for the entire family. It was a success for Scarlett; she was able to gather enough food to last two months.

However, Hunter thought otherwise. "My ribs are clutching my skin, and I am hungry out of my mind. I don't think I can take this any longer."

"Can't you wait a few minutes? The food isn't ready yet."

"Well, I'm not waiting for the food because I don't have an appetite," said Hunter.

"What? Why's that? You and Ma gathered enough food for all of us. Why don't you want anything to eat?" asked Carmen.

"I don't know about you, but I'm not used to seeing my meal alive before I eat it. I saw Mom hunt everything we're about to eat. And the worst thing about it is that she tried to make me help her skin all those animals all by myself after she slaughtered it," said Hunter.

"Yeah, she's always been into doing things like that. It's what attracted Dad to her. She was raised by a family of hunters, and out of all the women in the city, she was only the one who was not just attractive but also could provide for her family. That's where your name came from. When you were born, everyone wanted you to become everything Scarlett's grandfather was, but obviously you're nothing like him," said Carmen.

"Fuck you! If you were there, you would've felt the same way. I know you couldn't handle it because I know you've got a weak stomach; that's why you can't suck Raymond's dick without hurling," said Hunter.

"Well, you'd better get used to this, you little prick, because your lips will never touch another silver spoon. We will never return to the city or even reach space," said Carmen.

"You don't know that. I believe in Dad. He will rebuild the spacecraft in time before the people reach us. But if they do reach us, the people of the city will learn the truth, and they will have a change of heart. Besides, Pops did a lot for everyone in that city. He built that city from nothing. There is no way the people can't come to an understanding once they find out he didn't do anything wrong. Unless he did something wrong," said Hunter.

"Do you think Dad did what the people think he did?" asked

Carmen.

"I don't know, but if he did what the people think he did,
then there's no turning back. I know the people of the city; they
don't take nothing lightly," said Hunter.

***

Audio Entry #07

"At sunset, the family and I began eating our feast created from what Scarlett and Hunter had gathered. It was beautiful, the entire family gathered around the campfire. However, during our total bliss from our current situation, of course Hunter had to break the mood."

Hunter had some questions for his father about their current situation.

"Dad?" asked Hunter.

"Yes, son? What's the matter?" I asked.

"How long do you think it's going to take you to fix the spacecraft, so we are able to leave Earth? I hate it here," said Hunter.

"It's going to be a while, son. There are a lot of things that need to be done before we can reach space. But understand that I am trying to get us to space," I said.

"Good, 'cause I hate it here," Hunter said.

"Think of it like this: although we don't have the things we had in the past, we're blessed to still be alive," I said.

"Blessed? We're *blessed*? That is something you tell a fool to believe. Blessed is having food on the table and a real roof over our heads. We have none of this shit! We're far from being blessed; we're fucked," said Hunter.

"Hunter, enough! This is not the time for us to be doing this! It's time for you to grow up, be a man, live up to your name, and weather this storm for your mother and your sisters. If you keep acting like this, we're going to fall with you because we are only as strong as our weakest link. And right now, you're look very weak. Man up! Got it?" I said.

"I got it."

"Good; now go to your cabin and think about what I just told you," I said.

Later, as the sun began to set, I was putting together the last part of the body frame of the Space Arc. Once I covered up the remainder of the spacecraft with a tarp, I called it a day.

The next day, Alchemist began the process of rebuilding the Space Arc. Scarlett made breakfast from the food that she'd gathered from the day before. From the embryo fluid inside one bird's egg, Alchemist made scrambled eggs. From the other egg's embryo, Alchemist made a balut. The supply of fresh water ran out, and the nearby river froze over, so they used filtered urine. Hunter tried to grab the skillet with his bare hands.

"You see, you burned your hands because your hands are too soft. You need to build some calluses on your hands. Don't worry; we'll work on that when we go hunting today."

Hunter sighed in frustration.

After Alchemist finished his meal, he went outside the spacecraft, where he would continue to rebuild. Alchemist had material that he was able to gather from Abe Smith's farm. Once he gathered what he needed, he carried it all outside. In one hand, he had a box kit with a welding gun and torch inside, and in the other hand, he had the material. Meanwhile, Alchemist and his family began to settle down. Everything was looking positive for the family.

Sade's vital signs and condition had improved significantly. As for Alchemist, he was making significant progress with the spacecraft.

However, what they did not know was that danger awaited them. Alchemist had started rebuilding the Space Arc by welding the body frame. Then he rebuilt the thermal protection system with shuttle tiles of materials such as ceramic quartz fabric and fiberglass. As Alchemist was working on the spaceship outside alone, a red Lotus approached the Space Arc with her cubs' meal in her mouth.

After he coated and glued the outside surface, he began rebuilding the propulsion system. The issue with the Space Arc's rocket engine was that the propellants, tank pumps, power head, rocket nozzle, and satellite had been shattered in the rocket blast. This was a difficult task; however, his mission was to rebuild the Space Arc's rocket engine to travel back to space.

As soon as the Lotus laid eyes on Alchemist, she dropped her cubs' meal and bolted toward where her cubs were resting. She seemed to feel that the Space Arc was her new home and Alchemist was the intruder; her cubs were in danger, so as she approached Alchemist, she didn't hold back.

Aero-9 synced itself into robot form and came outside to assist Alchemist in fixing the Space Arc. As Alchemist laid his material down and got ready to open his tool kit with the welding guns and torch inside, the Lotus charged Alchemist. She clawed and pushed Aero-9 out of the way, ripping its right hand off, and then attacked Alchemist like a cheetah. As soon as Alchemist heard something creeping behind him, he backpedaled and stumbled. The Lotus tried to grab him, but he kept his distance.

The tool box fell to the ground as the Lotus ripped through Alchemist's skin and flesh horizontally. As Alchemist lay on the ground wounded, the Lotus stood above to intimidate him. Alchemist crawled backward. The creature walked toward him slowly. Blood mixed with saliva covered the snow as blood flowed from the laceration in his leg.

Meanwhile, Scarlett heard the commotion and stepped outside. As soon as she saw what was going on, she grabbed the welding gun and torch from the ground and torched the Lotus to death. The Lotus was now dead; however, Alchemist had suffered serious gashes to his left leg. He needed serious medical attention before it was too late.

***

In the meantime, while traveling to his mother's grave, Mars decided to work on his fighting skills. In his private hyperbaric chamber inside his aircraft, Mars was able to set up a workout plan with the AI interface. He created obstacles and scenarios to fight through. First, Mars began his stretch. Next, he started with basic isometric exercises. After he finished his circuit training, he began his sparring session with androids. The scenario was to destroy all six androids; however, the obstacles were from the bottom of the deepest sea. In the depth of the challenge, Mars swam his way up while fighting androids equipped with highly advanced AI and weapons from a micro-minigun and dazzling laser eyes. The androids shot bullets and lasers; however, Mars was able to deflect, counter, and defeat all six androids with his bare hands. The stress and pressure began to build his muscles. Mars was supplied with a small amount of oxygen, but still he managed to escape and go ashore.

After Mars finished training, he began the process of recovery. As he was soaking in a hot tub, he realized that the temperature was dropping, so he decided to hunt for an animal that he could use for its fur. As soon as he stepped out, Mars saw a blue cheetah nearby. It looked like the cheetah had migrated up north on its own and was looking for something to eat. After Mars chased it down on foot, with his bare hands, Mars was able to kill the cheetah and skin its blue fur for a coat. Mars had exerted a lot of energy and now needed something to eat, so he ate its flesh raw from the bones.

After Zeno's conversation with James, he found himself in a tough situation. With both his sons on the space watch for indictment, Zeno found himself in a position where his children, his creations, were accused of terrible crimes against humanity, and essentially, as the leader of the United World, he had to choose his nation over his family to preserve his reputation as an honorable person. His position as the leader of the United World, along with his image and reputation, was in jeopardy if he didn't step in and do something about it. He could lose his freedom if found guilty of the crime of obstruction.

However, he'd found the solution to this problem. When he created Mars, he'd implanted a microchip in his lower back in case one-day Mars wandered off and was lost. This microchip would come in handy, because now he could track Mars down before anyone else found him. As soon as he returned to the executive palace, he began searching for the microchip to find Mars. When he found the microchip number, he inserted it into his tracking device.

Zeno found the exact location of Mars, and he began packing all the essential food and weaponry he would need. Zeno knew if the

UWSM reached Mars first, he would be mistreated and would suffer more. He knew that Mars wasn't mentally strong enough to survive in prison, so he needed to get to him before it was too late. Once Zeno was ready, he left the International Space Station to travel to Planet Earth.

Today was the day that Apollo would give his speech in which he would implement the new laws of the land; however, he was feeling nervous and tense. Lucky for Apollo, Zyana had the remedy to calm his nerves.

"Are you ready?" asked Zyana.

"I don't know. I am kind of nervous. I don't know how they are going to react when I present them these laws," said Apollo.

"Come here," said Zyana.

Zyana grabbed Apollo by his shirt, turned his back to the throne, and pushed him onto it. She got on her knees and began to caress his thighs as she looked deep into his eyes. She pulled his pants and underwear to his ankles. Once his penis was exposed, she first caressed and then engulfed his testicles. Apollo leaned his back toward the edge of his seat as she began to masturbate him and give him fellatio. She used her tongue to push his penis onto the roof of her mouth. Drops of her saliva began to fall from her mouth and flow down his penis onto his testicles and then onto the throne. She continued until Apollo climaxed in her mouth. Quickly, Zyana rose from her knees and then sat on his lap, where she tongue-kissed

Apollo. She then whispered in his ear, "Now you're ready. Go kill it,

baby."

Later that day, after Calvin and his group had finished their tour and settled down in their rooms, Calvin entered the soup kitchen to get something to eat and to find his daughter and wife. It was a scramble to find them in the crowd of people, but he did. He met his family for the first time since leaving for war. His daughter, Skylar, and his wife, Grace, had already gotten settled at their dining table near the end of the cafeteria next to the window. He approached their table. Once Grace realized that Calvin had returned from war, she greeted and embraced him before introducing him to his daughter for the first time.

"Calvin? Calvin! Ohhh! Oh god! What are you doing here?" asked Grace, hugging him.

"I am home; they sent me back home," said Calvin.

"Oh my God; I didn't know. When did you get back?" asked Grace.

"I came back today. As soon as I arrived at the space station, I came here to try to find you guys," said Calvin.

"Oh my God; that is so great! Do you know how long you can stay?"

"I don't know yet. All I know is that part of the president's legislation was to send back over a hundred thousand troops to the space station, and I was one of them. With the number of years, I served, there is good chance it'll be permanent, unless the army needs me," said Calvin.

"Well, since you are here right now, you can meet Skylar," said Grace.

Calvin had been at war when Skylar was born, so he didn't know his daughter; the only time he'd ever had a conversation with her was when she was an infant. Whenever they had the opportunity, Grace made sure Skylar kept in touch with Calvin. When Skylar was an infant, Calvin would have hologram calls with his daughter. But things changed when Calvin and the army of space marshals moved along the galaxy. As the distance grew between father, daughter, and wife, the connection dissipated; he was light-years away, and there was no way he could get any satellite connection. So, in Calvin's mind, Skylar was still a baby, but, she was no longer an infant. Her voice, physical stature, and facial features had changed. Skylar was seven years old and more conscious and aware of everything that was going on in her life since the last time he'd had a hologram conversation with her and her mother. Calvin could tell this by how she approached him maturely and introduced herself.

"Hello, my name is Skylar. Nice to meet you."

"I know who you are, Sunny."

She blushed, because at that point when she approached him, Skylar approached her father as a stranger. But when Calvin referred to her as Sunny, she knew he knew who she was because only people who truly knew her called her by her nickname. However, she didn't know that her father was the one who had given her that nickname because of her curly brunette afro, big bright smile, and her wide bright hazel eyes. The only father figure who had ever been present in her life was her grandfather. When Calvin went to war, his grandparents, his wife, and his child came together to support and help one another while he was away. When Skylar started to talk, the only person she called papa was her great-grandfather.

Grace had taken offense and had had a discussion with Skylar's great-grandparents about not allowing her to call her great grandmother "mom" or great-grandfather "dad." "I feel it's important for her to know who her real mother and father are. I respect everything you've done for us, but I think it's important that Skylar knows who her father is, so she is not confused."

"Fair enough. I didn't think anything of it, but I'll make sure Skylar knows who her parents are. I believe that's as important as you do," said Skylar's grandfather.

At that point, whenever Skylar called her great-grandparents "mom" and "dad," they always corrected her and told her stories about her father, the type of person he was, and the amazing things he had accomplished. That way, when she finally met him and stood in front of her father, she'd feel she was in the presence of someone special. Whenever his parents had the chance to mention Calvin, they always made sure that she knew who her father was. However, Calvin was still a stranger because they had built him up to the point where he was an urban legend to her. That's why when Calvin reached his hand out to greet Skylar and called her by her nickname, she ran behind her mother, but when Grace told her it was OK and explained to her who he was, she opened a little more to her father.

"Skylar, what's wrong?" asked Grace.

"I'm scared."

"Aw. Why are you scared? That's your dad," said Grace.

"Are you my dad?"

"Yes, I am your dad."

***

Later that day, Dallas was interviewed, and the interview was broadcast live at Cyber News headquarters with the famous news anchor Michelle Waters. Michelle introduced Dallas and asked a series of questions.

"According to the *Daily Reporter*, unemployment for humans is at an all-time high at fifty-two percent. Due to the increased usage of robots and artificial intelligence, there is a decreased need for humans in the workforce. However, the secretary of labor and employment for the United World James, Dillion-Hall, is in the process of creating a reform that will increase the participation of humans in the workforce. When Secretary James was asked about how his reform would help increase the involvement of humans in the workforce, his response was, although there is no longer a need for humans in the labor force, it is important that we create opportunities where individuals are able to provide for not only themselves but their family as well. The gap between the rich the poor has increased. However, once this reform is approved, hopefully we can change the condition for everyone living in the

International Space Station.' There will be a meeting next Tuesday on his reform. We will keep you posted as this story develops.

"In other news, OXYGEN X4 stocks are down two percent, and Blue Gold is up ten percent. Synthetic water went up from its original price of ninety-five dollars to one hundred dollars, while organic water is still ten thousand dollars. Star Energy's stock value has gone down eleven percent since the last quarter. The chairman of the CHNOPS Company Star Energy, Olivia Ayers, expects to see a rise in the stock as the company goes through new managerial changes…

"Speaking of Ayers, I am here today sitting with Dallas Ayers, the son of Zeno Ayers and the brother of Mars and Alchemist. Dallas is a child prodigy who grew up on the International Space Station. The fifteen-year-old is an engineer and a research scientist, and he is also an entrepreneur who mastered and surpassed great physicians, physicists, biologists, and philosophers, and who practiced bio electromagnetics. He is the CEO of AERO, Inc., and this golden child is estimated to be worth over a hundred million dollars.

"He has won the Nobel Prize and the Albert Einstein World Award of Science. So, Dallas, what brings you here today? asked Michelle Waters."

"I am here to clear my family's name. I want to erase all the false information that was said about them earlier on this show." said Dallas.

"Before I ask you the obvious questions, tell us about yourself. What inspired you to get into the field of science and engineering? asked Michelle Waters.

"Well, it all started with Alchemist. I was inspired to be like my older brother Alchemist. Although I was very young, when I had the chance, I would read my brother's notes and books, and I was able to learn about advanced mathematics, modern sciences, engineering, and technology, and I mastered them all. I became fascinated with the power of science and what you can create with it. As I began to get older, my brother started to guide me, and I learned more about the power of science. I went from a fascinated boy to one who saw the potential of science to save the world. When creating my personal robot Pilot, I built him to accommodate my needs for assistance and protection. However, I realized that I could help other people who may have similar issues, and I shared my creation with the universe. My robot Pilot can help anyone—a doctor or someone who is sick and needs assistance with living. Pilot is built with the capability to use human logic in different scenarios and the ability to analyze problems or understand human emotions. That's what my brother Alchemist taught me: use your power for the good of humankind. He's the one who gave me the idea to create AERO, Inc. Unlike other scientists, Alchemist taught me how to use science to create change and peace among everyone in the universe. What Alchemist wants out of life is to bring peace to the universe." said Dallas.

"But didn't your father use science to plague Planet X the same way your brothers, Alchemist and Mars, did on Earth? Your father deceived the people of Planet X, and your brothers followed in your father's footsteps. using the same battle tactics in which he used to plagues The City of Neoterra, said Michelle Waters.

"It wasn't Alchemist or my father. It was Mars." said Dallas. Michelle Waters was shocked. She wasn't expecting that remark. She was baffled and didn't know what to say, so she quickly cut the interview to commercial once again.

***

After Iris took the thirty-four gods' powers away and dethroned them as gods, he outlined his new laws.

"Today I stand among you all as a god. I discovered the original hidden verse and that I am indeed the one and only god. No longer will you worship Queen Nebula. I am the Lord, the Alpha God. I am your shepherd. I am king of the universe. You will worship me! Nothing will be done without me being present. I will control your souls. I will control your deepest desires. When I command you to kill for me or covet for me, I will be right next to you as you walk through the fire.

"On Planet X, women were dominant and the noblest individuals in society. The gods were spirits, and Queen Nebula was the only god who could appear before other life-forms. She was the symbol for women, the supreme breed. The Amazons were warriors who were chosen by the gods to protect the well-being of the universe and to protect the forsaken land called the city of Utopia; however, they could only be women. Men were subject to pleasing women. Every day us Titans had to clean the Amazons' bodies, lactate the Amazons' breasts, sexually please and feed the Amazons.

However, now that I am Alpha of the universe, the Amazons' role on Planet X has been reduced to slave role.

"As for the Amazons, no longer will the Amazons be held as supreme beings. The Titans will rise again, and as of today, you serve and worship us. Amazons, you will learn how to please Titans, worship the ground we step on, and serve us accordingly. All rise, Titans. All rise, Titans!"

Once Alpha Iris finished his speech, he rallied his new Titan army to invade, capture, and imprison all the planetary gods in an eternal bliss asylum: Aquila, Cetus, Hercules, Cepheus, Hydra, Cygnus, Vulpecula, Lyra, Gemini, Draco, Aquarius, Perseus, Ursa Major, Ophiuchus, Monoceros, Lynx, Camelopardalis, Carina, Sagittarius, Scorpius, Vela, Puppis, Lupus, Ara, Musca, Norma, Hydra, Centaurus, Caeli, Columbae, Draconis, Phoenix, Taurus, and Auriga.

***

In the Space Arc, Sade was resting in a state of confusion. As she lay in her bed, she could hear her father crying for help and fighting for his life. The sympathy in her heart and concern for her father's well-being caused her to cry as well. When Carmen heard her little sister crying, she came to her aid to comfort her and calm her down. "Sade, what's wrong?" asked Carmen.

"Dada," said Sade as she pointed toward the wall.

"Aw, Dada is going to be OK. OK? Don't worry. Everything is going to be all right."

Carmen checked her vital signs, and everything was fine. She covered Sade's ear with headphones and played some soothing music to calm her down and help her fall asleep. Meanwhile, Alchemist was in another room bellowing in pain. Hunter and Scarlett were also in the room trying to aid him. Scarlett treated the wound; however, his leg was badly damaged. After Scarlett closed Alchemist's wounds with a hemostat and then cleaned the wound area, he lay in the bed in agony as he fought for his life. Hunter stood in the room, helpless, because there was nothing he could do other than stand there and watch his mother try to heal his father. Once Scarlett finished doing what she could do for Alchemist, he was then placed in a hyperbaric chamber where his leg would be examined by Aero-9. Aero-9 and its high artificial intelligence was able to diagnose that, during the attack, he'd also suffered severe nerve damage, and now his leg was infected and needed further treatment. Alchemist was given hyperbaric oxygen therapy to help improve his condition.

Meanwhile, as Alchemist continued his healing process, Scarlett received a holographic message from someone. She stepped into the other room where Sade was resting to answer the call.

It was Mars, who wanted to know what the status of the family was.

"Yes, how are you doing, Mars? It's Scarlett."

"Hey, Scarlett, how's Sade? Is she dead?"

"Well, not really—"

"What you mean 'not really'? What's happening?"

"I gave her the poison like you told me, and the poison had its effect. However, Alchemist noticed something was wrong, so he quickly gave her treatment and healed her."

"Where is she?"

"She's right here. In this room."

"Let me see her."

"Do you want me to do anything?"

"No, not right now. I want you stay put. I want you to act normal and make sure no one knows you know anything. Meanwhile, I will be traveling to Earth. Once I arrive, I will tell you what needs to be done next. Coordinate your location on Aero-9's system, and I will be there as soon as possible. Got it?"

"Got it."

"All right; I'll see you later."

"Bye, Mars. I love you."

"I love you, too."

Then Mars hung up…

***

After the smoke cleared, once again, the people of the city gathered in the courtyard of Alchemist's palace. After Apollo approached the balcony that stood above the crowd of people, he set foot on the balcony, and he began his speech.

"Citizens, gather around. As you know, today I will be giving you all the new law of the land. First, I am giving social power to you, the people. As of today, I no longer should be referred to as Professor Apollo. Today I am known as King Apollo, and you will refer to yourselves as kings and queens. I still will lead you all in the right direction if needed. However, in the end, no one man should have all the power; we should share our power equally—that is why we are all kings and queens now. No longer will there be a divide between women and men, black and white, rich and poor. From this day forward, all are equal.

"Second, cash is king no more! Currency brings envy and division. I declare the city a free society. This will eliminate poverty and wealth. When Alchemist was the leader of the city, he made sure everything was conducted in a capitalist fashion. But now we are turning this city into a currency-free society where there's no need for money. This will eliminate all job specialization, all social classes, and all mediums of exchange among the people in the city. To the homeless and hungry, I will provide a home with all essentials at no cost. As the new leader of the city, I will use the same philosophy as the Native Americans: we will live off the land. Women are in control of the land, and men control the distribution of goods from the land. And once those goods are ripe and ready, they are considered the city's property, and everyone in the city will share them equally. Everything that grows on the land or lives on the land is yours. We are all equal now!

"No longer will you be bound to one woman or one man. No longer will you be forced to obey one religion. You can live where you want to live and practice your own religion. You have free will. However, among our people, we do not condone killing or putting anyone in harm's way unless it's Alchemist or someone who is loyal to him. Alchemist is our only enemy."

The crowd cheered.

"And finally, no longer will we refer to this city as the city of Metropolis. It is now the New Apollo."

The crowd cheered.

After everything began to settle for Apollo, he began to act on all his promises to the people of New Apollo. The first area Apollo focused on was rebuilding the infrastructure of the city. Apollo began to rebuild the streets, buildings, and homes for impoverished families and individuals. Apollo made sure everyone had all the essentials to sustain societal living conditions without worrying about being unable to afford the basics. The last thing he did was structure a new centralized government. Apollo hired men he could trust. From the Alchemist Academy, Apollo created his own executive branch of power. His cabinet included a commander in chief of agriculture, a commander in chief of defense, a commander in chief of education, a commander in chief of housing and urban development, and a chancellor.

***

The Davis family quickly reunited with one another and tried to catch up; however, this was short lived, as Calvin was called for his interview with the shelter director.

"Calvin?"

"Yes," said Calvin.

"Hello. My name is Shelly Smith. I am the shelter director. I will be conducting your entry interview. How are you?" asked Shelly.

"I'm doing well," said Calvin.

"Good. Take a seat. As you know, I will be asking you a few questions about your current situation to help you figure out how I can assist you in getting you back on your feet. So how did you find the homeless shelter?"

"I recently returned from battle."

"How long have you served in the space army?"

"I served over seven years in the space force."

"Oh, wow, seven years. Is it only you?"

"No, it's my wife, Grace, and daughter, Skylar, right now. My daughter is seven years old."

"Wow. You have been in the service as long as your daughter has been around; is this your first time meeting her?"

"Yes, it's the first time I've met her. I had hologram conversations with both Skylar and Grace years ago, but it was never enough. When Skylar was first born, I would keep in contact with them, but as she got older, it became more difficult to keep in contact as the distance between us grew. Plus, after the death of my mother in-law, we couldn't maintain communication. Therefore, when I returned to the space station, they didn't know I'd been sent back home. When I first arrived, I was searching for my family back here in Station 13. I didn't know where they were at first; I searched in countless shelters. As you know, there are over a hundred homeless shelters in Station 13; plus, they never stayed in one shelter, so I had to look everywhere. It wasn't until my former family pastor was able to direct me here. I visited my pastor to see whether he had seen my daughter and my wife or knew where they were staying, but he wasn't sure where they were. However, luckily for me, he hinted that there was a good chance they were staying here, and he was right," said Calvin.

"That's wonderful. How is it to return to the space station?"

"Now that I'm back and the family is all together, the challenge I have to face is reconnecting with Skylar and Grace. It's hard because when I left for battle, I was a child. But when I returned, I returned as a man who didn't understand how to function in the modern world. I understood how to protect and serve this nation, but I didn't know how to be nurturing or knew how to provide for my daughter and my wife yet. For Grace, my physical appearance and mentality have changed. I wasn't sure if Grace could embrace me. I knew I was still attracted to Grace, but I had doubts whether Grace would love me the same way. It's been seven years since I last saw Grace face-to-face. I'm a different man. I wasn't sure how my changes would affect our relationship. When I met Grace, I was physically fit, and she was attracted to me. But now I am different. As you can see, I lost my arm, both legs, and a portion of my face, and I suffered genital wounds due to the war. We replaced all those parts with prosthetics, but I lost confidence in myself. When I look in the mirror, I don't feel human anymore. I feel like half man, half robot. I wasn't sure if Skylar or Grace would embrace me as I wanted them to do. But lucky for me, I married a woman who is a great person and has truly unconditional love for me, and she is raising a child who shares her values. I was afraid that Grace would

be mad at me, the way my mother felt about my father when he left

the family. I thought Skylar would hate me as I hated my parents

when I was sent to live in an orphanage; I understand how it feels

when a parent leaves for war. My entire childhood and adolescence, I

lived with my grandparents. When I was the same age as Skylar, my

father went into the force to fight in the space war. He was one of

the top lieutenants for the United World. He was one of the first

groups of human beings to reach Planet X. However, he was killed in

the line of duty. As for my mother, she abandoned the family and

gave me away for adoption when I was nine years old because she

couldn't deal with my father not being there to support her and

raising me as a child alone."

"Why are you leaving your son?" asked Angelina.

"I'm just not fit to raise him anymore," said Miri.

"I know exactly what you are doing—you're using your son as a tool to get back at my son, Calvin, but I can promise you, it's not going to work," said Angelina.

"How can I love a man and take care of his son when he loves his nation more? If he cared about me, why would he leave me and my son to live by ourselves?" asked Miri.

"Where is my grandson? If you don't want him, give him to me. That's my grandson," said Angelina.

"I gave him away," said Miri.

"Where?" asked Angelina.

"Social services. I called them, and they came and took him away. I don't know where he is right now," said Miri.

"You are going to regret everything you're trying to do to my family. You're trying to destroy my family, but I will stop you. I suggest you leave right now before I do or say something that I may regret," said Angelina.

"Fine. I'm leaving," said Miri

"What was wrong with your mother?"

"She was diagnosed with bipolar disorder and wasn't stable enough to raise me as a child. For a couple of years, I was an orphan, until my grandparents on my father's side finally found me, and I was adopted by my grandparents when I was eleven years old and I lived with them until I left to join the space force in 3000, the same year Grace gave birth to Skylar. Grace had conflicts with her mother because of the pregnancy. Grace's mother didn't approve of our relationship, so when she was three months pregnant, Grace was forced to leave their home. She moved in with me and my grandparents. I left when Grace was six months pregnant. My grandparents took in my wife and child and supported them while I was away fighting until my grandfather died in 3002 and my grandmother died in 3003. Having two other adults in a stable household was important, not only for Skylar, but also for Grace and me as well because it eased the pain of separation and helped them cope. It was like when my parents left the family, and my grandparents understood that my daughter and wife were in the same predicament that my mother and I were forced to live in. They understood what they needed and how to meet those needs.

"This is why Grace has been faithful to me—my grandparents taught Grace how to live without me and raise our daughter without a father. I can tell Skylar loved my grandparents because she has a photo of them in her necklace locket, and she holds that dearly," said Calvin.

"So, what happened to your daughter and wife after your grandparents passed away? Is this when things changed for you and your family?" asked Shelly.

"They were forced to fend for themselves. Grace maintained the household after the death of my grandmother, but a few years after their passing, they had to live with her mother because she couldn't keep up with household expenses. After losing their home and her mother in a house fire, they have been homeless for the last three years. When I returned to the space station, I knew I would be forced to face new challenges, but this time it would be with my wife and daughter," said Calvin.

"But you would be putting your daughter in the same predicament that you were in as a child, so why join the space army when you could have done something else?" asked Shelly.

"When I decided to join the space army, my grandparents didn't feel that I owed the war anything, but I felt obligated to fight in the war for my father to avenge his death. This vacancy in my life truly affected my relationship with my daughter and my wife. I was searching for closure. Every day when I stepped on that battlefield, I hope was hoping to meet the man who killed my father face-to-face. Maybe I did, or maybe that person met his own demise, but luckily for me, I didn't die trying to find out whether I did or not. As I am starting to get to know Skylar, I realized that she's older and wiser than I expected. As I begin talking to her more and more, I can see she is starting to embrace me more and more as if she'd known me all her life. Coming back from the war has been difficult because of the anxiety that built up inside of me, but I know my family will make my transition easier."

"That's good to hear. Well, now that I know a little about you and your life, and since you are a veteran and have family here, I'll make sure that I come up with the best plan to help you and your family get back on your feet."

***

Back at the Cyber News studio, Michelle resumed the interview where they'd left off, when Dallas revealed that Mars was involved in plaguing the city. However, before they went back on the air, Michelle wanted to know if he was still willing to reveal his bombshell. "Wow. I wasn't expecting you to reveal that information to me. This is big news. Are you sure you are going to go through with this?"

"Yeah, I'm sure; that's why I'm here—to clear my family's name."

"Interesting. All right. Are you ready to get this started?"

"Yes, I'm ready."

"All right, and rolling at five, four, three, two, one," said the producer.

In the second part of the interview, Dallas continued talking about his relationship with Mars and what he believed was the motive.

"So, what is your relationship with Mars?

"There is no relationship. People always assume that Mars is my brother, but I am here to tell the people of the United World and the rest of the universe that he is not. Mars is Alchemist's clone. Mars was forced into the family. My father created him for an experiment he put together when I was younger."

"So, Mars is not Alchemist's twin brother, and in fact, he is a cloned version of him?

"Yes, and there is a huge difference between being a clone and twin. If you ever met two people who are twins, the personalities are sort of the same; however, when you look at Mars and Alchemist, they're not the same. Mars has a personality of his own. I believe that Mars felt like he was the outsider of the family and wanted to sabotage the family's name because he felt excluded. Mars is jealous and hates me and Alchemist because our father gives us more attention and treats us better. That's why he's trying to kill Alchemist and my nieces and nephew: to stop that from happening.

"That's another question that I have for you. Is Alchemist alive? Some people say he's dead.

"No, he's alive. That's why I will be traveling to space—to search for Alchemist and his family and bring them back to the International Space Station."

"Well, thank you for coming here today and clearing your family's name. I really appreciate you for coming."

***

Later, Adakis's mother, Alexandria, was able to gather enough information to help Adakis so he could get his father's respect back.

"All right; it's confirmed. Pluto indeed sent Mars on a mission to find and kill Alchemist. This is your chance to redeem yourself and prove you are worthy to be the leader of your father's empire. Alchemist is on Planet Earth. You need to find Alchemist. Make sure you get him before Mars does. When you find him, kill him, and then bring his body back here. Get the ship ready, and in the meantime, I'll send you the coordinates."

"All right, Ma."

"And make sure you don't bring that boy Angel. You know how I feel about him. I don't trust him."

CHAPTER THREE

Episode 12: Love Lost

***

Although Calvin had been in tougher circumstances in his life, in his heart he felt that he wasn't ready to be a father, husband, or even a regular citizen. Calvin was still adjusting to regular life outside of war. And getting used to expecting the unexpected was hard for Calvin. Being that Calvin was in the army for almost seven years, he'd grown accustomed to the surprises that war brought. He had seen it all as far as war was concerned. But when it came time to adjusting to regular life, it was tough for him. When he first learned that he would be returning home, he was looking forward to adapting to the daily routines the citizens of the International Space Station were blessed with. But once he realized what he had to face, he learned it would be a huge adjustment. When Calvin arrived at the homeless shelter, it wasn't his ideal homecoming. His family was homeless and living in the shelter. Plus, he had a daughter and wife whom he could not provide for on his own. This was a problem for Calvin. It was like he was on a battlefield but this time fighting a battle that he was not prepared for. Living in the homeless shelter brought a new perspective to Calvin. Making the adjustment from leading an army of men to becoming the protector and provider of a household brought fear into Calvin's heart because war was all Calvin knew. However just like war, he would learn that the next day was not promised. He didn't know what tomorrow would bring, and this was the case last night. Every day Calvin was paired with someone different. He didn't know whether if the man next to him would was on drugs or he would overdose on drugs and die in his sleep. A man who was sleeping in the same room with him had died from an overdose. After seeing them carry his body from the room to the ambulance, Calvin was traumatized. Seeing a dead body brought him to a bad place. It reminded him of war, and at that point he wanted to leave the shelter; however, he had to deal with it for the sake of his family—plus he had nowhere else to go.

From that point on, when Calvin rested his head on his pillow in that homeless shelter, he thought about never being able to get on his feet and his daughter and his wife being subjected to the same problems that other homeless people faced for the rest of their lives. It scared him. It hurt his pride that he couldn't provide for his family. Eating, sleeping, and showering with a hundred strangers in cold water really started to get to him. At that point, he was determined to change their situation. So, Calvin spent the next few months searching for a job.

With Shelly's help, he looked for work. Calvin had no job experience other than fighting in the war. The military was all he knew, so he decided to change the type of job he was applying for. His grandfather had been a police officer for thirty years, and he wanted to follow in his footsteps. Calvin contacted the United World Police Department. He applied to become a space police officer. Once his application was approved, to move forward in the process, he had to pass all the required training, provide his medical records, and pass the physical exam. Calvin submitted his information and passed the training, but his mental health was questioned.

Although he had prosthetic body parts and suffered major war wounds, this was the lesser concern, as he had the strength of twelve men. But the amount of brain trauma he suffered while fighting in the war was a problem. Calvin didn't show signs of any mental disorder, but while fighting in the war, he had suffered over two hundred brain traumas. The space force worked Calvin like a machine, and the police department didn't think he would be fit for the police force and denied him a position.

"Your history shows that you are a highly respected soldier. You were a part of historical battles and led highly classified missions. With the things you have accomplished for the army, you would be the ideal candidate. But given the type and severity of the injuries you suffered, I am sorry to inform you that we are unable to hire you. There is a huge difference between the police force and the army. I have been a part of the police force for close to a decade. Although I've never met anyone who's accomplished as much as you, I've seen soldiers who have joined the police force. I've hired ex-soldiers, but a lot of them can't tell the difference between war and a tight situation. See, the difference between cops and soldiers is the level of power that you are given. As a soldier, your job is to meet fire with fire. As a police officer, you must protect the people but also serve and interact with them. You would not have the same power to use deadly force unless they posed a threat to you or others. I want to thank you for your interest in joining the police force. We respect everything you have done for this nation, but I can't take the risk because of the medical record. I'm so sorry," said the Station 13 police lieutenant.

Later that day, Calvin returned to the shelter. He was wondering what to tell his wife and daughter about what he was going to do next, now that he'd lost the police job. But when he arrived, he was greeted by his old general from the war.

"Calvin, how are you?" asked Space Marshal #48201E-95.

"General, what are you doing here?"

"I am here for you! We want you to return to the space force. I need you to lead a group of space marshals on a highly classified mission. I know you just came back and you have been reunited with your family after seven years, but we believe you're the only one who can lead this group of men. Come back to the army, and do what you do best," said the general.

Coming back to the homeless shelter after hearing what the police lieutenant had to say about his health and knowing that he may not be able to provide for his wife and child, Calvin needed something that could stroke his ego. So, when he made his decision, he quickly chose to return to the war.

"I'll go, but you have to be the one to explain to my family that I'm leaving them again," said Calvin.

"All right. That's fair. I will do that for you," said Space Marshal #48201E-95.

Before Calvin had arrived at the shelter, the general knew his family was there but hadn't approached them. It wasn't until Calvin pointed in their direction that he finally approached them. As he got closer, he was trying to figure out how he would explain to them that Calvin would be returning to the war. Space Marshal #48201E-95 picked up Skylar and began to break the news to the family. In the beginning of the conversation, Grace smiled, but as the general marshal began to break the news, her facial expression dropped in disappointment.

"Hello, what's your name?" said Space Marshal #48201E-95.

"Skylar."

"Hello, Skylar, my name is Space Marshal #48201E-95. How are you?"

"I'm fine."

"I am a general for the space army, here to bring your father back with me. Let me ask you a question. Do you know what your father does?"

"I don't know," said Skylar.

"Well, he's a space soldier, and his job is to keep the bad guys away from you, your mommy, and everyone else in this room. Your dad took a break from fighting, but we need him to come back and help us win this war because he is a very good soldier. Is it OK if we bring your father with us?" said Space Marshal #48201E-95.

"It's OK."

"All right, thank you. I promise we will bring him back," said Space Marshal #48201E-95.

Throughout the conversation, the general never acknowledged Grace. He never made eye contact or asked her permission to take her husband back to the battlefield. He made it simple and short and sweet. After he put Skylar down, he walked past her and said, "I am sorry, ma'am. You have a nice day."

Once Space Marshal #48201E-95 walked away from Skylar and Grace, Calvin approached his family one last time before he returned to war. Calvin wanted to return to the war and lead a special-mission unit to help him capture Mars and Alchemist. At first Calvin was reluctant, but once he looked at the situation that was going on back home, he felt that he had no choice but take this mission and leave his family to fend for themselves once again. When he told his wife, she gave him the choice: fighting a war or being a husband and a father.

"Honey, I'm sorry..." said Calvin.

"I waited for you for seven years to come back. You come back for a few months, and suddenly, this man comes here, and now you're going to leave us again? Don't leave. Please! Don't leave! Your daughter needs you! I need you," said Grace.

"I'm sorry, but I have to do this. Believe me, I don't want to go, but at this moment, I feel like this is the only purpose I have in life. I am trying to provide for you and Skylar, but no one seems to be willing to give me a chance!" said Calvin.

"You didn't get the job?" said Grace.

"They feel that I am not fit to join the police force. I'm done. I have no choice but to go back to war," said Calvin.

"Then keep fighting here with me and Skylar! You're no good over there. You've paid your dues. You don't owe anyone anything. Stay with us, and fight with us!" said Grace.

"I understand what you're saying, but this is my choice. I thought I was going to have peace of mind if I would ever come back home, but I realize there is nothing here for me. At least when I am on that battlefield, I know that my efforts are appreciated," said Calvin.

"What are you—a soldier or a husband and father? Skylar and I have been homeless for three years, and you're still leaving us? If you want to use this war as a way of escaping your problems, go ahead. But understand that if you leave us, we will not be waiting for you. I don't think I can wait for you any longer."

"That's fair…but before I leave, can I say good-bye to Skylar?"

"I think it's best if you don't. She doesn't need to remember when her father ran away from her…"

"That's fair. Just tell her I love her. I love you, too," said Calvin.

"Calvin, are you ready? All right; let's go," said Space Marshal #48201E-95.

"I guess this is good-bye," said Calvin.

"You be safe," said Grace.

"I will…"

Calvin waved good-bye to Grace, and then he was on his way to fight the war.

***

Back in the Space Arc, Alchemist was recovering from his flesh wounds. As Alchemist rested in the recovery chamber, he learned that the worst part of being attacked by a Lotus was surviving that attack. As days passed and the temperature dropped, Alchemist's muscles, ligaments, and tendons were tightening up, and his bones was aching. Aero-9 examined his leg to see whether there was any improvement in his condition.

"AERO, how does my leg look?

"How do you feel?" AERO asked.

"I feel better than I felt yesterday. Is it raining?"

"Yes, and it's quite gloomy."

"I can feel the barometric pressure on my leg; it feels like an elephant is sitting on me."

"On a scale from one to ten, how serious is the pain?"

"Ten. My whole leg hurts."

"Well, according to the impact analysis, I was able to estimate the severity of the injuries and how long your healing process will be. Things are not looking too good. Your knee was shattered. You have ruptured your kneecap and torn the muscles and ligaments, and you have a tendon rupture in your left leg. After I repaired the ruptured muscle and torn ligament in your quadriceps region and knee, I made some progress on your leg. But there is an infection spreading to the deeper tissues. To treat the infection and speed up the healing, I will need to increase the dosage of hypersonic therapy. Once we see some progress, we'll go from there."

As Aero-9 lifted Alchemist's leg, its hand gave out.

"AERO, are you OK?" Alchemist asked.

"Yeah, I'm fine. It's just my hand. I was able to repair the hand that was damaged during the attack, but it's not at full strength. But once I recharge my hand, it should be fine," said Aero-9.

"Sounds good. Thanks, AERO. Keep doing what you're doing.

I don't know what I would do without you."

"No problem. That's what I'm here for."

In the meantime, with their father injured and Hunter being the only son in the family, for the family to survive and eventually reach outer space, they'd have to rely on the leadership of Scarlett and Hunter. With Scarlett by his side, Hunter was being put to the test to become the man who his mother and father always wanted him to be. Hunter was being introduced to adulthood from adolescence in the wilderness. He found himself trying to prove his toughness and his manliness to his parents, although Hunter didn't believe he was worthy of taking on that role. A high-spirited young man, he had a lot of confidence in himself, but he didn't have the tendency that his father or his mother wanted him to have.

Alchemist and Scarlett had high expectations for Hunter. Alchemist came from a family of wealth and accomplishments. The Ayer family had made huge contributions to the advancement of society. Zeno, Alchemist's father, had created the first interstellar ark and later established the first branches of government, a space bank, and business ventures such as space mining. As for Scarlett, she came from a family who served in the army on Earth after the apocalypse. Scarlett's father fought for and served the people of zone three before Alchemist's arrival. He fought in multiple extraterrestrial invasions and earned medals of honor. And when he returned home from war, he taught what he did in war to all his children. Scarlett's father, Anderson, always preached toughness and the importance of standing for something. Anderson was family oriented and wanted his eleven children to be able to hold their own but protect and provide for their families as well.

Scarlett was raised in zone three in Metropolis, the poorest section in the city. Before she married Alchemist, Scarlett lived with ten siblings and her mother and father. Scarlett was raised very harshly by her parents, but she was raised with discipline, unlike Carmen or Hunter. When she was home, her mother, Lia Auburn, taught her how to be nurturing to her younger siblings. But her father, Anderson Auburn II taught Scarlett to be a provider and protector for her family. After her young brother, Nicholas, was old enough to follow his lead, Scarlett's father taught them how to provide for the family by teaching them advanced hunting and fishing skills. Scarlett's father also showed her and Nicholas how to protect themselves and the family by teaching them the skills he used when he was in the military. Scarlett's father taught her some advanced hand-to-hand combat, archery marksmanship, and survival skills. Therefore, there was friction in the relationships with Scarlett and her children.

When Scarlett met and married Alchemist, it was a huge adjustment in her lifestyle. She had worked for everything, but Carmen and Hunter had everything handed to them. Being that Hunter was the only son, Scarlett wanted him to be more of man than a boy. When Scarlett and Alchemist came together and had a son, their expectation for him once he got older was that he would be the best of both families. They thought that he would have the heart of Alexander the Great but the mind of Albert Einstein; however, Hunter fell short of their expectations. Alchemist wanted his son to be able to take the leadership role, but Hunter was not built to be a leader. Like most teenagers, Hunter had a free spirit. He didn't look at things as broadly as people like his father did, and he didn't worry about serious things like people in positions of power had to. The issue that created the strain with Scarlett and Hunter was that sometimes his demeanor appeared weak. He was not living up to his names—Hunter or Ayers. And she believed Hunter appeared weak because of how he was raised.

She felt that living in The City of Neoterra had been a disadvantage for him; although Hunter was the secretary of defense, he'd never tested his survival or battle skills. As a leader of an army, Hunter had experience in different elements in combat, but now he found himself in an element that wasn't familiar or comfortable territory. Hunter was a highly motivated teenager, but the challenge that he was forced to face now was honing the necessary skills to survive in the wilderness, which he didn't have. He found himself struggling to survive. Prior to the family's departure, the city had been in a state of peace, and Hunter was shielded from hardships in the utopia that his father had built. The population of humans on Earth was over a million. In that region, there were only about thirty thousand, and in The City of Neoterra, only about twenty thousand people. Other than the people in the city itself, there weren't any threats. The skill Hunter learned from being the secretary of defense was policing the people of the city, not engaging in full on combat. Now he was forced to face the realities of the real world. Death was his reality.

The first time Scarlett and Hunter went out and hunted, Hunter watched and learned. But the next time, everything was hands on. Before Scarlett went out with Hunter into the wilderness, she set targets, so he could practice his aim. When she put him to the test, things didn't go well at first. Hunter shot, and the arrow barely reached the target. So, Scarlett stood behind Hunter and wrapped her arm around his chest and arm to teach him the proper form.

"I'm not going to give you the fish; I'm going to teach you how to catch the fish. I'm going to give you a fishing pole. If I teach you how to eat on your own today, you'll be able to feed yourself for the rest of your life. First thing first: change your position. You must make sure you're pulling with your dominant hand. Keep your feet shoulder-width apart. Next, stand straight up and relax your body; you don't want to be stiff and tense. When you are pulling the arrow, pull the string with your middle and index fingers, and clutch it with your thumb. Point at the target, and release the arrow," said Scarlett.

Scarlett handed Hunter the bow and arrow, and he prepared to hit the target. But he dropped it when a field mouse appeared. In her frustration, Scarlett grabbed the bow and arrow from the ground, quickly aimed, and killed the mouse in its tracks. She then handed the bow and arrow to Hunter, so he could again try to hit the target.

"Go now, do what I showed you. *Now.*"

***

Finally, Zeno reached Planet Earth. Zeno had Mars on his radar, but before he could land on a flat surface, he'd lost the signal. Without a signal, Zeno couldn't use the tracking device to find Mars. The only information he could use was the coordinates from where he was last located. Considering how large Earth was and the distance Mars could travel in a matter of minutes, the task of trying to find Mars would be tough for anyone. Zeno calculated that Mars was traveling in the northern region of the Americas at about 550 miles per hour, so he knew he was not traveling on foot but most likely in an aircraft or a spacecraft. Zeno had no clue where Mars might be, but if he had to guess, he might be heading to the grave of his mother, Bella.

Growing up, Mars had been very close to Bella. Although Mars wasn't her biological son, they had a very close bond with each other. Mars was a very emotional individual; when he got attached to something, he held it close to his heart. So, when she died of cancer, he took it hard. She was buried in Weston, Massachusetts, where she was born and raised. When Mars was younger, Zeno and the family would sometimes take a trip to Earth to visit her grave and pay their respects. And every time they visited, he didn't want to leave. It was a struggle to bring him back home. Sometimes they had to drag him onto the spacecraft. And sometimes he would leave and travel back to Earth on his own and sleep there for days before someone found him. If Mars was indeed traveling toward the New England region, the only place he would go in that area was to Bella's grave.

***

It was April of 2991. Spacecraft swarmed around the International Space Station. They all were a part of Elea nation,[10] and they were preparing to abduct human life-forms. Once the command was given by King Tabbris Elea, the invasion ensued.

A large group of humans were abducted by the Elea nation and transferred to Planet Europa. They were going to serve as slaves for the Omegan people. However, from the large group of people, Omega hoped to find his future queen.

"Make sure the humans are transported to the plantation. I want them stripped of their clothing and working as soon as possible. Halt! Stop that line. Excuse me, miss. Hello, I am King Tabbris, king of the Elea nation. Welcome to Planet Europa. What is your name?"

"My name is Ozelia."

"Ozelia. My, what a beautiful name. I couldn't help but notice you standing in line with these worthless creatures. I was wondering if you would like to come with me to my palace."

"Sure, I guess."

---

[10] The Elea nation is the land of the Martians and Angel and Tabbris Elea family.

It was Ozelia's unique beauty that made Tabbris pick her from the crowd. From that moment, Tabbris fell in love with Ozelia. They soon lived would together as king and queen.

***

Soon after the new rules were set, the citizens of the New Apollo began making their adjustments to the changes in the city. Now that the New Apollo was a free society, no one had to work, which meant nothing would be done. At first, everyone was enjoying themselves. They had access to all the food, water, alcohol, and so forth. Everyone partied and wasted everything for thirty days straight. Everything was good for everyone in the city, but once the supplies ran out, things began to become very difficult. People began to starve, and soon people began to rebel. The Mason family was one of many that began their new life in the New Apollo. However, unlike the other citizens of the new Apollo, the Mason family had a lot more resources that would help them later. Although there were individuals who had the skills to restore the condition of the people, they didn't have the resources to restore the condition because the people wasted everything away. As for the Mason family, after they picked an area in which they wanted to live on, they began to build upon the land. With just a couple of biologically advanced planting seeds that they supplied themselves with from space, they were able to create their own crops and then reuse the same seeds. However, they only had so many seeds. It was enough to feed the Mason family but not enough to feed an entire city. This brought jealousy and envy among those in the city.

Meanwhile as everyone in the city was trying to figure out where their next meal will come from, Apollo was sitting on his throne. Across the room, Zyana was teasing him sexually and slowly undressing. She began to give him a lap dance as he caressed her. He pulled her breasts from her bra and engulfed them, and then he reached his hand inside her underwear and fingered her until she climaxed. During their foreplay, Apollo's secretary came into the room to inform him that a building he'd finished rehabbing had collapsed and multiple people were injured. Apollo quickly got dressed and went to see what was going on. He realized he was starting to lose control of what was going on in the city. In the first months as leader, he had experienced several other unexpected events. Although his new laws created true equality, there had still been a rise in violent crimes since he'd been in office. He heard a knock again.

"What is it now?"
"It's the Mason family. They're being attacked. What do you want to do?"

"I'll be right there."

Apollo walked all the way from his palace to the village where the Mason family was being attacked.

"Give us your food! You come on our planet, and you think you can just come here and build upon our land without paying your dues."

"What's going on?"

"Me and my family haven't eaten in days, and these space monkeys wouldn't give us their food. Give us some of your god damn food, you greedy bastards!"

"No; this is our food. I grew it with my own bare hands. I am not going to give away something that I grew for my family."

"Enough! Go to my palace; I have some food there. Bring your family. When you are greeted by a guard, tell them I sent you all there."

"Thank you, King Apollo."

"As for you, Mason family, what you did was wrong. You know the rules—everything that grows on the land or lives on the land is for everyone to share. Next time when someone asks for some of your food, you give them food—understood?"

"Yes, sir."

"Thank you."

At that moment, Apollo was thinking whether he made the right decision in changing the rules of the city. However, he couldn't look back; he had to die by his decision…

***

Back on Planet X, Alpha Iris had started his building project in which he would use the human slaves to destroy the old temple and statues of the thirty-six planetary gods and build massive luxury buildings and statues of himself. In the meantime, while Alpha Iris's men began the construction of those buildings and statues, Alpha Iris was partying, drinking, and orchestrating orgies in the forbidden temple with the other Titans, breeding with humans and goddesses and Amazons. Prior to Queen Nebula leaving Planet X, the Titans, Amazons, goddesses, and humans weren't allowed to follow the acts of mortals, but now that she was gone, Alpha Iris and his coalition were acting out without fear of punishment.

"Have you ever made love with a god?" asked Alpha Iris. "Today is the day."

Iris grabbed a female by the waist, turned her over, grabbed her hair, and penetrated her anally as Sirus watched in disgust. He began to second-guess the morality of what Alpha Iris was doing. Guilt filled his heart, so he removed himself from the group, and in another temple, he began to pray to the gods.

"I come to you as one of your faithful worshippers. Lord, I ask for forgiveness. Forgive Iris and everyone who is partaking in these blasphemous actions. I fear that you will wreak vengeance upon us. Lord, please give me guidance; I don't know what to do. I feel guilty because I could have done something to stop this from happening. But I fear what I'm capable of. I think Iris has lost his mind. I pray you take away his power before he abuses it and someone innocent gets hurt."

***

At this point in Hunter's young life, he had a lot to lose. Danger was imminent and becoming the leader and the provider for his family was his reality. Hunter could feel the pressure falling on his shoulders, but in the process of trying to find his way, he found himself more concerned with defending his masculinity and not losing his parents' respect than with the danger of the wilderness.

"In the animal kingdom, there are two types of creatures: prey and predators. Which one are you?" asked Scarlett.

"I'm a lover, not a fighter, but I'll learn how to kill to survive," said Hunter.

"Would you kill for love?" asked Scarlett.

"If there is one whom I love, I will kill for love."

When Hunter talked about love, Scarlett knew he was talking about Carolina Summers. Back home, the only thing that concerned Hunter was impressing women. One who constantly occupied his mind was Carolina Summers. Carolina Summers was a twenty-year old woman who worked as a secretary in the city hall in Metropolis. Hunter was the secretary of defense for The City of Neoterra, so every day he tried to figure out ways to impress her. But she always turned him down. However, a few months before Hunter was forced to leave the city, she started taking him seriously. The relationship evolved into a romantic one, and they made love.

However, the relationship came to an end when the family was forced to leave the city. Although they no longer were able to see each other, Hunter still thought about her. He wanted to be with her, but it was impossible. But once he'd shown his lust for Carolina, Scarlett quickly questioned his love for her.

"See, when you have nothing to live for, you wouldn't do what's necessary to survive. That's the reason why the prey outnumbers the predators. They simply don't have anything to fight for. The reason the lion is the king of the animal kingdom is not because it's the strongest creature in the animal kingdom; it's because the lion knows what's necessary to survive. One of the weakest creatures in the animal kingdom is the gazelle. Gazelles travel in herds, and they look like they're all together, but they're not. They just travel together because if a pride of lions attacks them, they use each other as distractions so the rest can get away. They scatter and make it every gazelle for itself. That's their mind-set. When lions attack, it's for the good of the pride.

"Simply put, when you are a part of a family, it is love that keeps you together. Without someone to share that love, there is no love at all. That's why you need something to fight for to keep yourself alive. Love can be the thing you fight for, but you must know who that is. So, are you fighting for Carmen, your father, and me, or for that bitch Carolina Summers? Tell me, because I need to know where your heart is," said Scarlett.

"Why do you think I'm thinking about Carolina?" asked Hunter.

"Because I know you. Back at home, all you talked about was Carolina Summers. You loved her more than you loved yourself."

"No, I mean, I love her, but I love you guys, too."

"See, I knew you were thinking about her. Stop thinking about Carolina and start worrying about what's in front of you. When we were home, it was cute, but now I need you to forget her. Soon me, your father, and your sisters need you to step up and become the leader of this family. Your father may not make it through what he's going through right now, so as his only son, we need to make sure you are ready. Carmen is very bright, and she is older than you, but she can't do what you can do. Right now, you haven't reached your full potential. But you will live up to your first name. We need you to keep the family together. If you don't, we are nothing but a herd trying to survive in this animal kingdom. Got it?"

"Got it."

"All right then; let's go, pretty boy. There's no one to impress now but me. Let me see what you've got."

At that moment, Scarlett was trying to push the right button by mentioning Carolina Summers and questioning his manhood. Only time would tell if this strategy would work.

***

Later that day, Zeno finally reached Bella's grave. He landed a few miles away and walked to the site, where he found Mars speaking over his mother's grave. Mars expressed his love for his mother but also spoke of his father and siblings' relationships.

"Hello, Mom. How are you doing? I really miss you. I can't stop thinking about you. I know you're probably asking what I'm doing here, but a lot has been on my mind, and I want to talk to you. I can't go too much into detail, but I can tell you that I betrayed my father. I took the side of his enemy, and now he has put me in a situation where I've been asked to do something that I may regret. I love my brothers and my father, but I've been asked to kill Alchemist and frame my father.

"I know Alchemist is your firstborn and Zeno is the reason why I'm alive, but this needs to be done. There isn't much more I can say, but I love you and ask for forgiveness before I do what I must do. All right, Mom; I got to go. I love you. Please forgive me."

***

Pluto called Mars, and they had a brief conversation on Alchemist's whereabouts as Zeno listened from afar.

"Have you been able to find Alchemist yet?"

"Not yet, Pluto."

"That's fine; I couldn't either, but I was able to calculate where he might be. According to my radar, he was last spotted in the High Peaks Wilderness Area. I will send to you the coordinates. Get there as soon as possible," said Pluto.

"All right, I'm on my way right now," said Mars.

Mars was ready to leave, but he was stopped by his father, Zeno, who by eavesdropping had learned that Mars was plotting against him and Alchemist. Mars's demeanor changed. He went from angry killer to a little boy.

"Dad, what are you doing here?"

"I was looking for you; I knew I would find you here. What have you been up to?"

"Nothing much."

"Don't lie to me. I was here the whole time. You're plotting against me and your brother. Why are you doing this?"

"It's because you never loved me. Forever I have been searching for your love, but you treat me as if I didn't exist. Everything that I wanted, you gave Dallas and Alchemist."

"Tell me, Mars, what is it that I have been giving to Alchemist and Mars that I haven't tried to give to you?"

"You never loved me the way you love Alchemist and Dallas. You isolated me. You neglected me. You never acknowledged me as your child—only as your creation. Why did you create me if you don't truly love me?"

"That's not true; I treat you the same way I treat Alchemist and Dallas."

"Then when you created me, why didn't you give me the mind of Dallas? Why didn't you give me the ambition of Alchemist? All I've ever wanted was what you gave your other sons, so I can be a king in my own empire, be respected, and be recognized and loved by others…and so I can make you proud of me."

"So, you tried to take everything that you think I gave them."

"Yes, and I don't regret it."

"Son, come with me. I have to show you something."

***

Adakis and Mars had both reached Earth but at different times. Mars arrived on Earth before Adakis, but Adakis was closer to Alchemist and his family because Mars went to visit his mother's grave. Pluto didn't know that his son trying to kill Alchemist, but Alexandria had the inside information on the mission. Now he was on the hunt to track down Alchemist as well. Neither Mars nor Adakis knew where Alchemist and his family were yet, so Adakis was told to stay put and wait for further instructions.

After he set up a tent and started a campfire, Adakis reached out to Angel in the Republic of Pluto to see if he had any information on Alchemist's whereabouts.

"What's up, Adakis? What's going on? I've looked for you all day. Where are you?"

"I'm on Planet Earth right now. I'm on a mission to kill Alchemist."

"OK. What's the matter? Do you need some help?"

"Is there any way you can help me find Alchemist? I'm in the area where he might be, but I am unable to track him down yet."

"Sure, I can help you track him down. The Elea nation has some satellites surrounding Planet Earth that they set up after the invasion.

I can gain access to a satellite from Planet Earth to help you locate Alchemist and his family."

"I knew I could count on you. Thanks, Angel."

"Anytime. Just wait for a few minutes, and I will send you their location."

***

Back at the International Space Station, Calvin was finally transported from the homeless shelter to the United World Space Marshal Army Base in Station 51, where he would prepare for the highly classified mission. But before Calvin began getting his men ready for battle, he needed to pass his physical examination.

"Hello, my name is Dr. Kelly. Please state your name and the branch of the military you're currently serving and for how long."

"My name is Calvin Davis, my Space Marshal number is 2339IE-5, and my branch is the space army. I've been serving as a United World space marshal for over seven years."

"And how many missions have you been involved in?"

"Around two hundred in one year," said Calvin.

"How long would an average tour last?"

"A typical tour of duty for soldiers is a couple of days. A lot were top secret, and I was the top soldier in the squad," said Calvin.

"And have you suffered any severe injuries?"

"Yes, I lost my right arm, a portion of my face, both my legs, and my penis and testicles."

"Do you remember how it all happened?"

"Yes, I lost my right arm and the right side of my face in a friendly crossfire from a rocket. As for my entire lower portion of my body, I lost that in a land mine explosion."

"And because of these traumatic events, have you been diagnosed with any conditions?"

"Surprisingly, I haven't been diagnosed with any psychological disorder or symptoms."

"Really? Have you ever experienced body ache or battle fatigue? Have you had brain injuries such as concussions?"

"Yes, but when I suffered those injuries, I was able to recover fairly fast. When I lost body parts and suffered nerve and bone damage, my flesh ached, but I haven't experienced any other symptoms."

"No restlessness, psychomotor deficiencies, jitteriness, confusion, nausea, or vomiting? No hypervigilance and so forth? Really? Because two hundred missions are a lot," said Dr. Kelly.

"Not really."

"How about your physical functioning? Were you able to adjust to your new body?"

"I am not going to lie; adjusting to my new body was difficult at first. Every time I lost a body part and it was replaced with a prosthetic part, it was like I was losing control of my body and someone else was taking control of it. At first, my body and my nerves weren't adjusting to the prosthetics, but once I began to train harder and became stronger, I finally was able to take control of my entire body. Growing up I thought that if I could become a robot, it would be cool; I would be superhuman. I would be strong. I could fly. I could do anything I wanted. But when I essentially was becoming a robot, it was degrading. The way doctors looked at my body, opened, removed things, operated on it, and told me how to use it—you can never get used to them operating and *building* your body. But I guess these are the costs of fighting a rich man's war."

"Is it OK if I examine your body?"

"Sure."

When he removed his shirt, she saw Calvin's wounds and prosthetic limbs. Dr. Kelly really admired his brawny arms, shoulders, and abs. His bronze skin was striking against his prosthetic limbs of titanium and steel, and he had a large distinctive tribal lion tattoo on his back. She then measured and weighed him.

"Six feet eleven and three hundred sixty pounds."

After she finished with the physical, she ordered a special suit and medication and antidote to help him function on Planet X before he went on his way.

***

After Scarlett's talking with Hunter, he was frustrated about a comment that his mother made. At first, Hunter was filled with anger and anxiety, which made him unsure of himself. He didn't know how to react. However, once Hunter calmed his nerves, his demeanor changed. When Hunter and Scarlett stepped into the wilderness, his mission was to prove his mother wrong. Scarlett handed him the bow and arrow, and his killer instincts, inherited from his mother, quickly kicked in. Hunter began to use his preexisting skills, which he never knew he had, to prove his toughness.

"Look. Do you see them? Vultures. If you follow the vultures, you'll find a free meal awaiting you. However, beware; there will be others lurking as well," said Scarlett.

Scarlett and Hunter found the body of a mountain goat, and they watched a group of vultures feasting on the corpse. Scarlett passed Hunter the bow and arrow.

"Here. Let's see what you got. Try to hit one of the vultures. This should be an easy kill; they are more focused on eating their feast than what's going on around them. Try to hit one now before it's too late."

Hunter got into position. First, he placed his feet shoulder width apart. Next, with a strong posture, Hunter grabbed the bow and arrow with a firm grip before releasing the arrow. In one shot, Hunter was able to hit one of the vultures before the others scattered.

"Good shot; good work."

Hunter was able to get his first kill and gather his first meal for the family. Scarlett was able to push the right buttons, because at that moment he was a totally different person. Before when Hunter went out to hunt, in his mind, he would always debate with himself whether he should release the arrow and kill the animal. However, now Hunter was reacting like a natural-born killer. From that point on, Hunter was able to carry on the confidence that he built from that one lucky shot. At the end of the hunting trip, Hunter's hunting skills had improved to the point where he was killing bighorn sheep and antelope from long distances. When they finished gathering enough food, Scarlett and Hunter found an area where they could skin the animals they killed so when they traveled back, they could carry less and bring home the most important part: the flesh.

The first time when Scarlett assigned Hunter to skin the animals they gathered, the sight, smell, and feel of the carcass caused him to vomit and pass out. But now he was able to cut and skin the flesh like a butcher. Once Hunter was able to prove to his mother that he could be a provider and protector, he earned her respect. At first, there was a lot of hesitation, but once he gathered himself together, he was removing the stomach from the animal with his bare hands without cringing.

***

It was the end of the day. They had just finished up skinning their last animal when a black bear appeared in the distance. Scarlett wanted to call it a day and retreat, but Hunter was so high on the success of what he'd been able to accomplish in the wilderness that he wanted to test his skill one more time. Hunter was reminded to stay silent, not to make sudden movements, and to shoot only to kill.

"'Cause if you miss its head, it won't go down easy, so don't miss."

Hunter missed. Instead of hitting the bear, he hit the tree next to it. This alerted the black bear. When the black bear spotted Hunter and Scarlett, it charged them. Hunter was shell-shocked and did not move as the bear charged them. Scarlett grabbed the bow and arrow from Hunter and stopped the bear by hitting it between its eyes.

At that moment, Hunter learned the number-one lesson in hunting: hunting at goodwill. Respect this game called hunting. Never get blinded by the thrill of the hunt; stay focused and take all you can before it's too late.

It was another successful day for Hunter and Scarlett. In that day alone, they were able to gather enough food to last another month. It was all in a day's work. Although their hunting trip ended with a mishap, Hunter was steadily becoming a man as he learned from his mistakes.

"In order to survive here, we've got to kill what we eat. So, you must learn how to kill. I know it's very scary, but you need to get over your fear. Fear is the key to survival. Fear will give you a reason to fight. If you use it wisely, it can be your tool to be alert. It can teach you how to be disciplined and cautious and stay aware of your environment. However, if you let fear consume you, then your fear will swallow you alive. What happened back there could have happened to anyone who was put in that situation. But you must understand when to swallow your pride and know your limits, when to fight and when to retreat. That's the number one rule in the art of hunting: know your prey but also know yourself as well."

***

After they finished hunting, Scarlett and Hunter found a water fall where they were able to clean their body off. Scarlett didn't hesitate to remove her clothing in front of Hunter. At first, Hunter didn't want to look in her direction; however, once she forced him to join her in the waterfall, he had no choice but to look.

"Boy, hurry up and get in this water; we don't have all day. We have to head back to the spacecraft before it gets dark."

In his mind Hunter felt ashamed that he was looking at his mother nude as she washed away all the blood and gore from her entire body. However, there was no denying that he was stunned by her beauty. Not only was Hunter amazed by her body but was also mesmerized by her body tattoos. It was the first time he'd seen tribal body tattoos. He knew she had tattoos; however, he never had the opportunity see it in its entirety. From her neck to ankles, her body was covered in tattoos. After Hunter and Scarlett washed off their entire body, they both did some self-grooming before they left to head back to the spacesuit. From the reflection of the water, Scarlett noticed that her pair of her French braids was frizzy. So, she decided to rebraid it. After she was done braiding her hair with a blade,

Scarlett showed Hunter how to cut his long hair.

"Come here. Your hair is a mess. Let me fix it for you," said Scarlett.

"OK."

With Hunter pushed on her chest, with a comb, Scarlett combed his hair upward and then cut enough hair on each side of his head until he had a ponytail fade. After she finished cutting his hair, she then braided the remainder of his hair.

"You look like you father when he was younger. Handsome." said Scarlett.

Once they finished grooming themselves, they returned to the spacecraft. When they returned to the Space Arc, they had a rude awakening when they found Sade fighting for her life. As Hunter and Scarlett entered the Space Arc, they could hear the chaos. Carmen was screaming as Alchemist and Aero-9 tried to revive Sade. Scarlett and Hunter heard Carmen crying for help, and they ran to the room to see what was going on. The poison was again in Sade's body; her vital signs had dropped, and now Sade was fighting for her life. Alchemist and Aero-9 tried their best to revive her.

"What's going on?"

"The methyl mercury that was previously in her system had come out of remission and caused her to fall into a coma. Her mercury levels are elevated to four thousand micrograms per liter, eighty times the toxic threshold. We didn't know anything was wrong until Carmen went into her room to check on her and found her unresponsive. While Sade was sleeping, she slipped into a coma. This is my fault. I didn't follow the precautions. I knew I should have monitored and done more evaluation after we thought we removed the toxicant from her body. That could have prevented this from happening," said Alchemist.

At that point, Alchemist's faith was being tested. His leg was injured, and his daughter was fighting for her life. Alchemist was able to find the strength to try to revive Sade, but she passed away from complications from the poisons in her body. Sade was pronounced dead at ten o'clock that night. A few hours later, they gave their grace.

"Lord, grant us the serenity to accept the things we cannot change, the courage to change the things we can, and the wisdom to know the difference. Help us find meaning in this terrible death. We all are broken by this loss. Please give us the strength to move forward. Let us pray that you accept our daughter and sibling's soul as she tries to pass through your pearly gates. Please grant her passage because her heart is pure. In your name we pray. Amen," said Alchemist.

After they prayed for her soul, Alchemist and the family laid her body inside a casket and stored it inside a room in the Space Arc, so they could preserve and release her body once they made it to the International Space Station.

***

Zeno knew that he had to kill Mars. Mars was a wanted man. Everyone wanted to get their hands on him for the reward or the recognition. There was no telling what they might do, so he had to take the matter into his own hands. Zeno didn't want his Mars to suffer. So instead of telling him he was going to kill him, he decided he was going to lure him to the top of the cliff, and as they watched the sunset, he was going to slit his throat. He didn't want Mars to see it coming. He wanted to keep it short and sweet, so Mars didn't feel any pain.

As they approached the cliff, Zeno and Mars were deep in conversation.

"Mars, you asked me to love you. I can do that, but will you do something for me?"

"What is it?"

"Can you be honest with me?"

"Sure."

"Let me ask you a question, Mars. Did you plague Alchemist's city?"

"If you want me to be honest—yes, I did."

"Who else was involved? Was Alchemist involved as well?"

"No, just me and some of my acquaintances; none of them you know."

"Why did you do it?"

"I wanted your attention. That's why I plagued Alchemist's city and then framed him. It wasn't because I hated you, Dallas, or Alchemist. I tried to hurt the family's name because I felt like your forgotten son. So, the only way I could get your attention was if I earned your respect and competed with my brothers."

Zeno was preparing to kill Mars, but as they were reaching the top of the cliff, his conscience bothered him, and so he confessed what he was going to do.

"I understand. Mars, let me tell you something. I love you dearly with all my heart. Since you have been honest with me, I must be honest with you. Mars, you are a wanted man. You and Alchemist are on space watch for the crimes that you have committed on Planet Earth. You have a lot of enemies, and they want to see you suffer for the crimes you have committed. I don't know what they might do to you if they get their hands on you, so I have no choice but to kill you before anyone can get to you."

At first Mars didn't understand. He was shocked, but when he realized the severity of the situation, Mars sincerely accepted his circumstances. "If I must die, I would rather by the hands of the man who made me than the hands of a stranger."

Mars prepared to die. They sat and watched the sunset. Zeno looked deep into the eyes he'd created. But before he could slit Mars's throat, he was stopped by a space marshal. Another group of space marshals had found Mars and Zeno and had come to bring Mars into custody.

"Lord, give me strength and forgiveness for what I am going to do."

"Zeno, let him go! If you kill Mars, I promise you, you will be arrested and spend the rest of your life behind bars. Don't do it."

Zeno was still holding Mars by the neck, but he let him go. Mars was arrested by the space marshal and shipped back to space where he would be prosecuted for his crimes.

***

At West Point in the United World Space Army Base, Calvin passed the physical exam and was set to lead a presentation in which he introduced the objective of the highly classified mission.

"Hi, I am Space Marshal #2339IE-5, and I will be explaining our first highly classified mission. I want to congratulate you all in advancing in rank and the academy. For a lot of you guys here, it's been a rough few years. I am here to tell you that it's going to get a lot worse. You thought you understood what was ahead of you, but you were naïve. But you withstood the physical and mental demands. You completed your basic army training and passed to the Space Marshal Academy, where you all learned about the code of military justice. You learned your combat role, and then you were assigned your first duty station. You spent a few months training for combat. You learned how to clean weapons and how to put on and use your space suit. You began your simulation training and were put to the test. You engaged in true combat. You learned how to use highly advanced space vehicles. Once you completed training, you were off on your first mission. However, the battlefield wasn't what you thought it would be. The pace was a lot faster, and you realized in a matter of seconds that you could lose your life. I need you all to understand what you're getting yourselves into because I want to make sure you are ready for what's ahead.

"The main objective of this mission is to capture Alchemist and Mars Ayers. These two high-profile individuals are part of the first family of the United World, but they are very dangerous. They are on the space watch for conspiracy to commit mass murder and other crimes. They pose a threat to society and need to be brought into custody.

"This task will be very difficult because our destination is Planet Earth. For centuries, due to the damage caused by numerous nuclear wars, Planet Earth was uninhabitable for humans. Other than Mars or Alchemist, no man dared to return. But today we're put into a situation where we are forced to take drastic measures. You were all chosen to be a part of this mission because you're highly trained and experienced individuals. However, I can assure you that it's not just a matter of how well you can use your weapon or how strong or intelligent you think you are. It's that you are willing to die for your nation without hesitation. We will be in unmarked territory, and we have to expect the unexpected."

***

At post fifty-five, after Calvin's introduction of the highly classified mission to his new group, they had open weight training. Calvin would then begin his individual training. Before Calvin began his workout regime, Dr. Kelly synced information in his brain to help his performance. With the use of brain-wave synchronization therapy, Calvin would be able to enhance his military knowledge and muscle memory. Dr. Kelly was able to sync elements of combat such as the sweet science of boxing and tactics in advanced war strategy. Next, Calvin was equipped with a state-of-the-art space suit. The helmet was equipped for reconnaissance, with x-ray, ultraviolet, and gamma-ray protection, and his bodysuit was equipped with titanium, tungsten, steel, and carbon fiber, or compression garments made of fiber optics material and a respirator mask.

Once he was dressed, Dr. Kelly inserted a chip to monitor his health in the artificial gravity-training room where he would begin his training. At the beginning of his workout regimen, he would train under Earth's gravity levels. However, as the training intensified, the atmosphere would change as well. The purpose of Calvin's workout regimen was to test his endurance and threshold for pain, but it also worked areas of his body such as his quads, chest, shoulders, and triceps. As he endured altered levels of the atmosphere and the accompanying conditions, his mind was his terror dome, but he was in control. The body was a temple, and Calvin built it to be a machine.

The way he took care of his body, whether it was through carefully monitoring his diet or following his training regime, Calvin created a foundation and built it from the ground up to the point where he was able to surpass levels far beyond what a normal human could withstand. The chamber in which he was training was built so he could change the gravity or temperature level and match any elements in the universe, and still his tolerance would remain the same. Whether the temperature was 200 degrees Fahrenheit or gravity matched the atmosphere on the moon, Calvin was able to transition without any of it affecting him. Calvin's space suit weighed close to two tons, but he was able to maintain a level of balance and strength. Calvin's body was able to quickly adjust to these elements.

Once he finished his workout plan, he began a sparring session with a couple of prototype pyros droids. These were not ordinary droids. They were built with some deadly weapons, and the

AI was synced with the brain waves, strategies, and skills of famous dictators and war leaders. Whether it was the mind of Sun Tzu or Napoleon, it was synced in the AI of the droid. But he was able to fight them off.

After he was finished with his workout, Calvin joined the rest of the space marshals at the base center as if he hadn't gone through one of the toughest workouts that any man ever went through without breaking a sweat. In the café, he ordered mocha latte with cream. As Calvin began to take sips of his latte, he was approached by an up-and-coming space marshal.

James introduced himself and began to fan out to Calvin as if he was a celebrity. But Calvin wasn't very responsive. Normally, Calvin didn't really get attached to other space marshals due to a fear of becoming connected with them just to lose them in battle and then being left to deal with survivor's guilt. But when he realized the type of person James was, Calvin began to open.

"Are you Space Marshal #2339IE-5? I've heard so much about you. It's an honor to be in your presence. My name is James Tanielu, but you can call me Tombo," said James.

Instead of saying Space Marshal #017758E-4, James had given him his government name, which wasn't allowed in the academy.

"What's your rank, young man?"

"I'm a junior officer. I just moved up after twenty-six months in service," said James.

"That's good."

"The real reason I wanted to speak to you is because I want to know if I can train with you; I want to pick the mind of greatness like you," said James.

"Sorry, kid. I just finished my workout. Maybe next time."

Calvin was getting ready to call it a day until he was challenged by another space marshal, and he needed James's assistance.

"Save your breath, James; you're wasting your time. This guy has a dick up his ass," said Alex William (Space Marshal #017758E-4).

"Excuse me, do you know who I am? I am a sergeant major, and I can get you discharged for your comment," said Calvin.

"I've heard so many stories about you, and everyone kisses your ass because of war stories told about you, but I'm not sure if any of it is true. I am skeptical. If you ask me, it was all hype," said Alex William.

Once Alex William made that remark, Calvin asked James to help him lace a pair of boxing gloves. He'd decided to take matters into his own hands. Calvin decided to challenge Alex to a boxing match. At first, it was only going to be Alex, Calvin, and James in the boxing-ring area, but when Calvin challenged the young space marshal and he accepted his challenge, word got out. The legendary Calvin Davis was going to showcase his legendary combat skills. And it was to be fifteen rounds in fifteen minutes. The excitement filled the room as Calvin shadowboxed in the corner, envisioning what he was going to do in the ring. But when he stepped in the ring, things didn't go as planned. In the first three rounds, Alex was able to showcase his hand speed. Calvin was right-handed, and Alex was a southpaw, but Calvin was leading with his left hand. This allowed him to throw punches from different angles. Alex was beating Calvin to the punch, but once he was able to calm his nerves, use footwork, jab with his right hand, and slip inside his left hand with precision and accuracy, Calvin was able to make the right adjustment to turn the fight around. Calvin started to counter most of the punches Alex was throwing.

At that point, Calvin was dominating the fight, but in the sixth round, he cut his eyebrow on a head-butt, and they began to trade punches. Calvin was bleeding profusely, but he threw a punch that cut Alex's eye. This worked in Calvin's favor because he was able to take control of the fight. The room filled with the aroma of iron from the amount of blood that they spilled, but Calvin was able to come out on top of the fight. Calvin knocked Alex out at the end of the tenth round. Calvin won and rightfully earned Alex's respect. When Alex was able to get back on his feet, they shook hands and went their separate ways.

It wasn't until Space Marshal #48201E-95 gave Calvin the power to assemble the special-mission units to help him capture Mars and Alchemist that they faced each other again. The first-person Calvin chose to be in his unit was Alex William (Space Marshal #017758E-4). Then he chose individuals he'd had experience with in battle and who were highly respected because of their experience and advanced skill sets. Individuals such as snipers Space Marshal #018645E-4 (Nate Jones) and Space Marshal #014332E-4 (Dennis Canty) joined him in the quest to find Alchemist and Mars. Although, Calvin James Tanielu had little experience, Calvin was willing to take him under his wing and teach him everything he knew along because Calvin had a special consideration for him as a person.

"Really? Oh my God. Thanks, sir. I won't let you down," said James.

"All right; let's go," said Calvin.

"All right; I'm coming," said James.

Soon after, they got vaccinations that consisted of a compound of nuclear radiation and steroids. Then they all were on a spacecraft to Earth.

Calvin and the space marshals boarded the aircraft and began their voyage through space to Earth in search for Mars. When they finally arrived on Earth, he gave a speech.

"Motherboard to Earth!"

"All right now, before we step out of this spacecraft, I'm going to be honest; I don't know what to expect, but I promise to lead you all and try my best to keep you all away from harm's way to my full capability. Just follow my lead, and I will do my best to bring us back safe and sound. Are you guys ready?"

"Yeah, we're ready," said everyone.

"All right; let's go."

In case someone in the family tried to get in contact with Mars or Alchemist, the United World Space Army was keeping its eye on the entire family. Calvin and the United World Space Army were able to gather enough information about where Mars might be from the recent activities of Zeno's spacecraft coordinates. Calvin estimated where Mars might be. But during tracking Zeno down, the space marshals were attacked by a group of Lotus'. As the space marshals slowly approached Zeno's spacecraft, a Lotus jumped on top of James, knocking his helmet off and biting his neck.

James was able to push it off, but soon after, he collapsed to the ground. Calvin and his men then slowly tried to retreat.

"What the hell is that?" asked Alex.

Through the lens on his helmet, Calvin was able to figure out what type of creature the Lotus was. According to Calvin's helmet interface, Lotus' were a group of mutant creatures. These creatures have developed poisonous scales around his entire body. Physically, their stature was that of a five- to ten-year-old child's. But they were very violent; they'd inherited predatory tendencies. Lotus' found thrill in causing evil, harm, and killing. They were small, but they were not intimidated.

If they had to fight a larger animal, Lotus' were able to use leverage and proper technique to gain the advantage. Most Lotus' were wild and couldn't be tamed, but the ones facing down Calvin and his men were far more advanced than the ordinary Lotus'. These hyperintelligent creatures were militant and had been trained by Mars. When Mars came to Earth on his own, he started an experiment in which he assembled a group of Lotus' that he bred, harvested, and trained. Most Lotus Creatures have DNA of a human. Mars was able to tap in to the Lotus" human intelligence and train them to follow his orders and protect him by any means. Lotus' use the fear factor; if anyone invades their territory or poses a threat, Lotus' will attack, capture, and eat them.

When James dropped his weapon, a Lotus grabbed it and prepared to shoot it. It then pointed the weapon toward Calvin and gestured to leave the area. At that point, Calvin had to swallow his pride and abort the mission. Calvin and his crew immediately retreated to the spacecraft, carrying James. James was poisoned and needed medical attention. Immediately, Calvin got to the space base and tried to find a way to save James.

"I can't move my body! What's going on? Ah, my skin!" said James.

"Mayday! Mayday! Mayday! We have a young cadet who has been attacked and poisoned by an unidentified creature. He's having difficulty breathing, he's paralyzed from the neck down, and his skin is melting off. I need help. Send me something before it's too late," said Calvin.

However, it was too late. James was dead from complications.

Thank you for reading

for more information, go

to

Dbentertainmentllc.com